MW01678510

"In the beginning God created the heavens and the earth."

Genesis 1:1

"The universe was spontaneously created out of nothing."

Stephen Hawking
Brief Answers to the Big Questions

Introduction

An Amazing Fact:

After a caterpillar forms its chrysalis, it begins the breathtaking process of transformation: Its worm-like body breaks down into a cellular soup and then reorganizes into the structure of a delicate butterfly. The two-week process involves the precise coordination of countless molecular mechanisms. The genetic and biochemical underpinnings of this fantastic transformation are a powerful reminder of the countless other amazing designs found in the natural world.

The astounding metamorphosis of a caterpillar into a butterfly is a phenomenon that defies naturalistic explanations. How could such a complex process arise through mere random mutations?

After all, if you found a Porsche 911 parked in your garage, you wouldn't assume that it had appeared there by chance. You would inherently know that the sports car had been built by skillful engineers and mechanics and that someone had driven it into your garage. Even if you couldn't explain how it got there or how it functions, you would know that everything about it—from its twin-turbo V6 engine to its iconic steering-wheel emblem—points to intelligent design.

Did you know that thousands of years ago, the Bible affirmed that the universe and all its complex structures had a beginning—long before modern science confirmed it?

> For by Him all things were created that are in heaven and that are on earth, visible and invisible. … And He is before all things, and in Him all things consist (Colossians 1:16, 17).

Many prominent thinkers and scientists throughout history mistakenly believed that the universe has always existed. For instance, the famous Greek philosopher Aristotle (384–322 BC) taught that the universe had no beginning. And as late as 1948, renowned scientists like Fred Hoyle published papers supporting that belief.

However, around the turn of the 20th century, scientists began finding evidence that our universe must have had a beginning. Some of Albert Einstein's contemporaries said that his general theory of relativity pointed to an expanding—or a contracting—universe, implicating a starting point from which everything arose. Einstein, himself believing in a static universe, initially rejected this idea, making use of a "cosmological constant" in his equations to balance such cosmic forces.

But then, in 1929, astronomer Edwin Hubble brought forward stunning observations from his telescope that changed the course of scientific thinking—including Einstein's. Hubble saw that the universe was expanding; galaxies were accelerating away from us in all directions. This implies that if time were to be reversed, all matter would eventually converge at a single point in space and time.

The universe had a beginning! It was big news to those who didn't believe what the Bible has been telling us all along. (Does this mean God created the universe using the Big Bang? See p. 50.)

We often hear that there's a huge rift between the Bible and science, a great controversy between creationism and empirical evidence. We say boldly that it simply isn't true. You've been lied to, purposefully, for reasons that may surprise you.

That's why in this magazine, we will explore the evidence for the origin of our world—and, yes, the universe. We'll take a close look at how the complexity and beauty of our natural surroundings are best explained by the existence of an omniscient, omnipotent Creator. We'll also examine the shortcomings of evolutionary theory and demonstrate how the evidence found in nature perfectly aligns with the Genesis account of creation.

The truth behind our origins is not just an abstract intellectual exercise; it has profound implications for understanding the purpose of human life—including yours. So, we invite you to embark on this journey of discovery as we explore the fascinating intricacies of our world, its origin, and the divine Craftsman behind it all.

As you read this magazine, remember the three most important questions you can ask: **1.** Where did I come from? **2.** What am I doing here? **3.** Where am I going? You cannot fully answer questions 2 and 3 unless you have the answer to question 1.

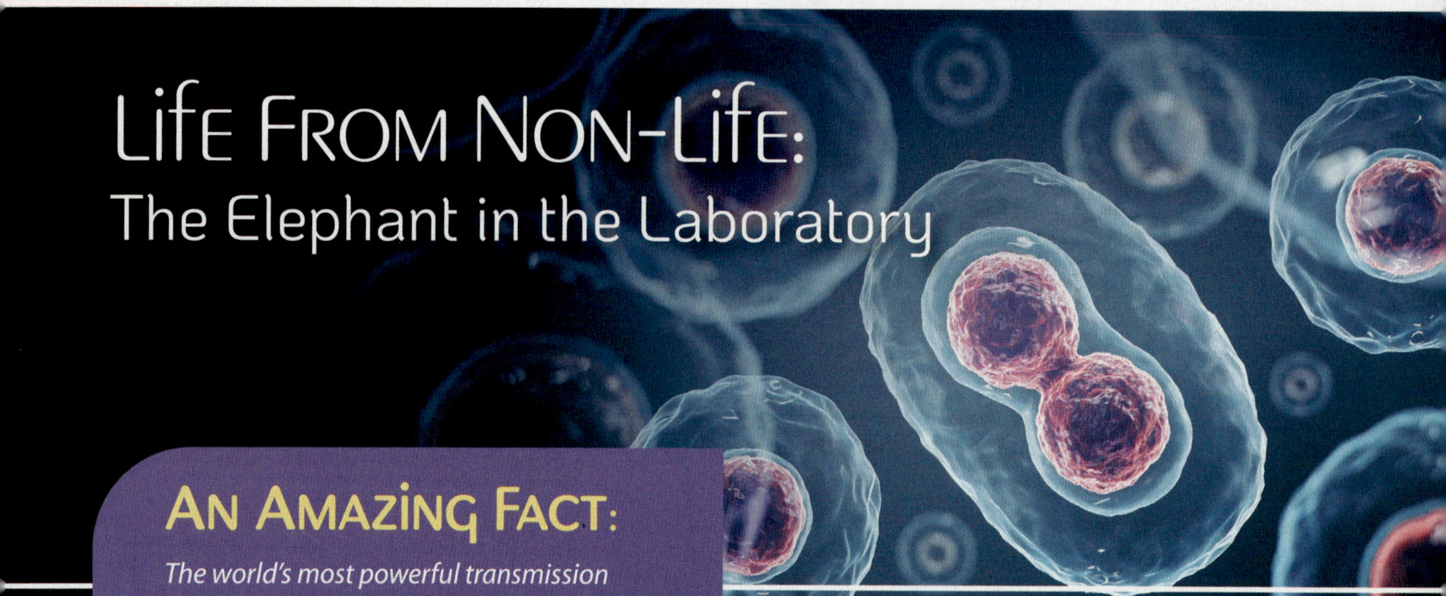

Life From Non-Life:
The Elephant in the Laboratory

An Amazing Fact:

The world's most powerful transmission electron microscope can magnify objects up to 50 million times. Located at the Lawrence Berkeley National Lab, the unit cost $27 million to build and uses 300,000 volts to operate. It fires electrons at nearly the speed of light, allowing it to focus on individual atoms.

©The Regents of the University of California, Lawrence Berkeley National Laboratory

Even though this planet is the only place where humans have observed life, it is teeming with flora and fauna. From invisible bacteria packed with DNA to massive blue whales plowing through the sea, our world is flooded with life!

But how did this incredible biodiversity originate? The evolutionary view is that life arose by utter coincidence from inorganic molecules. Building blocks—such as amino acids—formed and combined to make complex polymers. Simple organisms suddenly emerged from this "primordial soup" a few billion years ago.

This process is known as **abiogenesis**, the belief that life arose from non-living materials. However, abiogenesis has never been observed in nature.

Indeed, there was a time when people believed that creatures like flies and frogs spontaneously emerged out of rotting meat or slimy ponds. Then, scientists didn't know about the profound complexity of cells that make up even the tiniest living things. But in 1859, Louis Pasteur demonstrated that microbes do not suddenly appear in sterile meat; instead, he observed that the small worms found in spoiled meat came from eggs laid by flies. His discovery helped in formulating the **law of biogenesis**, which states that only living organisms can produce other living organisms.

Russian chemist Alexander Oparin soon challenged biogenesis by suggesting that organic chemicals could have come together to form the first living cell. In 1953, Oparin's theory was tested by other chemists who mixed several gases in an electrical chamber to create amino acids, the building blocks of proteins. Yet, those experiments did not create living matter. In fact, the destruction, not the

Panspermia

Another origin-of-life theory, called panspermia, proposes that life was seeded on Earth by intelligent aliens or by space dust or meteoroids. However, this theory merely kicks the can down the road, leaving the question: How did *that* life start?

Which Genesis?

Biogenesis: From *bio*, which means "life," and *genesis*, which means "beginning." That is, living things are produced by other living things.

Abiogenesis: The supposed transformation of inanimate matter into living matter.

production, of life's building blocks was observed to be the norm. Life coming from non-life is the proverbial elephant in the laboratory.

Scientists may use state-of-the-art labs to produce some of the raw ingredients of life, but from there, it would take a quantum leap to produce a living cell with sophisticated DNA code. With today's powerful microscopes, microbiologists know that even tiny life forms are packed with highly complex, interdependent systems that require intricate mechanisms to function, heal, and procreate. What are the chances of even one "simple" cell evolving randomly to life—not to mention surviving?

To this day, only biogenesis has been observed in nature and in the lab.

Why do some scientists passionately defend the idea that life randomly arose from non-life? How much more logical and attractive to believe the Bible's alternative! A wise and powerful Creator, beyond our realm of understanding, made life for a purpose because He is the embodiment of love. The Scriptures identify that Creator as Jesus Christ:

> In the beginning was the Word, and the Word was with God, and the Word was God. He was in the beginning with God. All things were made through Him, and without Him nothing was made that was made. In Him was life, and the life was the light of men (John 1:1–4).

Creationists do not have all the answers, but it is entirely rational to believe that the superabundance of complex life on Earth points to a supernatural and intelligent Creator.

Amazing Diversity of Life

One of the most powerful yet simplest arguments for divine creation is the mind-blowing diversity of incredibly intricate life on this planet. Astronomers haven't seen even a shred of life on any other planet in our solar system, yet Earth is exploding with a nearly limitless assortment of life. One would think that if life on this planet evolved from some elementary ooze where most of the creatures were subject to the same environment and ingredients, there would be only a few species—but there are over 8 million species of animals and 400,000 species of plants! (And keep in mind, those are just estimates because scientists believe they haven't found them all.)

The world is replete with animals that hop, skip, and crawl over land and animals that fly through the air. Some live underground—and some birds stay in the air for months without landing on the ground. Every animal species is highly specialized to fill their specific niche. For example, the giraffe's long neck is perfectly suited to its environment, where it's needed to reach high leaves. And the honeybee's sophisticated "waggle dance" communicates the precise location of food sources to other hive members. Additionally, plants conveniently convert sunlight into energy—supplying food for the animals. They blossom in every color. Vines climb and seeds helicopter through the air.

The vast diversity of life and the purposeful arrangements within ecosystems are evidence of a creatively and intentionally designed system. It is incomprehensible that all of the boundless variations of color and extremely complex interwoven relationships in these birds and bugs and animals and plants would have all sprung magically from the same brown primordial bubbling sludge. Natural selection and random mutations would not lead to such a rich tapestry of life. Believing that takes much more faith than believing an infinitely intelligent God designed it. "Now ask the beasts, and they will teach you; and the birds of the air, and they will tell you; or speak to the earth, and it will teach you; and the fish of the sea will explain to you. Who among all these does not know that the hand of the Lord has done this" (Job 12:7–9).

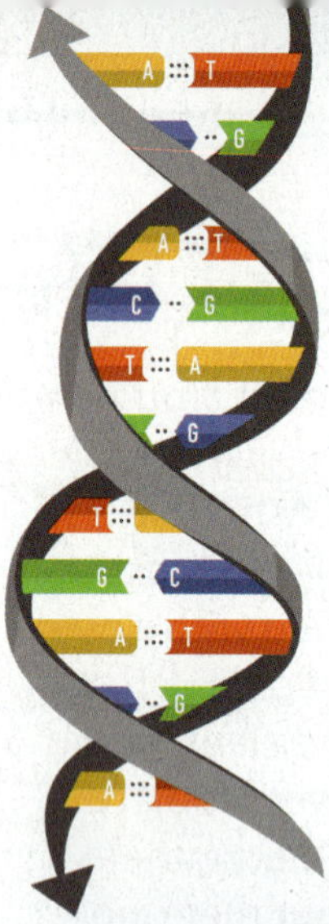

DNA:
The Code of Life

GUANINE
ADENINE
CYTOSINE
THYMINE

An Amazing Fact:

Mandarin is considered the most difficult language in the world. While more than a billion people speak it, Mandarin is challenging to master because it is a tonal language—one in which vocal pitch alone can alter a word's meaning. And with over 50,000 distinct characters, just looking up Chinese words in the dictionary takes a lot of skill.

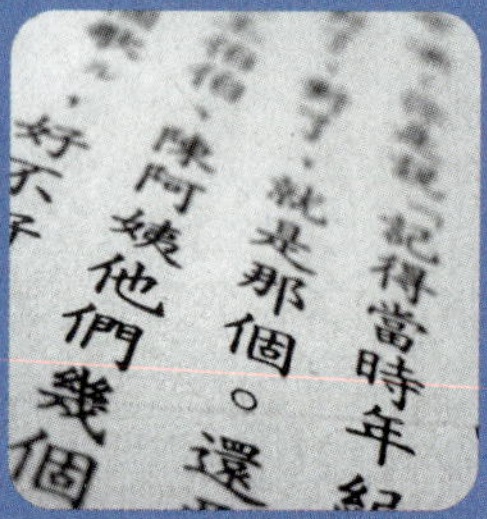

Today, every molecular biologist acknowledges that an even more complex language than Mandarin can be found in every living cell—the language of our genetic code. Ever since Francis Crick and James Watson discovered the DNA double helix in 1953, the secrets of this code have been slowly unlocked, leading to only one plausible explanation for its origins.

The complex language of DNA is based on four chemical bases that direct the ordering of 20 different kinds of amino acids, which form the many diverse proteins used to build our bodies. Thousands of different types of protein are manufactured as new cells are produced. As each unique cell is made in the human body, it replicates around three billion base pairs of DNA to make one copy of the human **genome**—an astonishing feat for just one cell!

A genome is a complete set of an organism's genetic material, including all its genes responsible for its structure and function.

Even more fascinating is the error-correcting systems found in DNA. With millions and millions of copies of code made with every new cell, errors do occur. While many of these are insignificant, an error rate of just one percent would devastate the organism. However, a built-in system works to correct

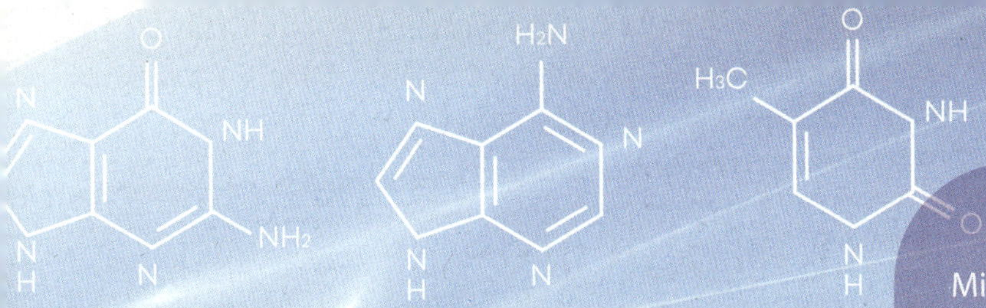

these errors through special enzymes that check and correct mistakes in the copy process.

How could this staggeringly complex system have evolved? Imagine being able to dial someone's 10-digit phone number after being given only two of the digits. How long would it take to guess the rest of them—in the correct order? Now multiply this by tens of millions of additional numbers, and you can begin to grasp the complexity of DNA.

The structure of DNA—its sophisticated language, mind-boggling replication, and error-correcting

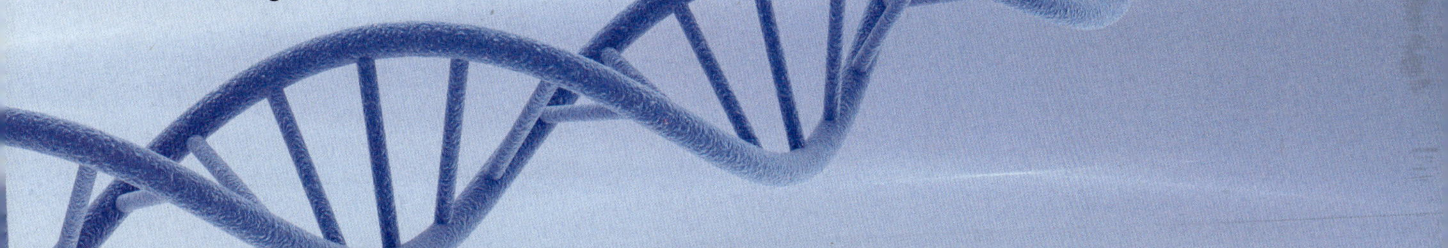

processes—raises serious questions for the evolutionary model. How could something prone to error allow for the evolution of a self-correcting system? How could such intelligent genes evolve? And how could a library of information be stored in one tiny cell? One researcher calls this difficulty "an unsolved problem in theoretical biology."

We can now clearly see the fantastically complex design and purpose in the tiniest forms of life. When King David contemplated the intricacies of living things, he wrote, "I will praise You, for I am fearfully and wonderfully made; marvelous are Your works, and that my soul knows very well" (Psalm 139:14). Surely, it takes a massive leap of faith to believe that the meticulous marvel and machinery of DNA could create itself.

Bill Gates, the founder of Microsoft, once said, "DNA is like a computer program but far, far more advanced than any software ever created." Yet, who would ever believe that his sophisticated Windows software created itself?

SOURCES

Roth, *Origins*, p. 138.

What About Junk DNA?

Scientists once claimed that vast segments of our genetic code are "noncoding," meaning they play no part in genetic expression; many viewed these as genetic leftovers from our supposed evolutionary past. However, further research has shown that much of this so-called "junk DNA" does, in fact, play specific and critical roles. Moreover, recent discoveries have revealed that a single strand of DNA can have multiple purposes and provide two kinds of information simultaneously.

The Cambrian Explosion:
An Unexpected Discovery

An Amazing Fact:

Tardigrades, or "water bears," are near-microscopic creatures with an astonishing ability to survive extreme conditions. Scientists were amazed to discover that these animals can withstand temperatures ranging from -458°F to 302°F, survive without water for 10 years, and even survive the vacuum of space. When an Israeli spacecraft crashed on the moon in 2019, it was carrying thousands of tardigrades—incredibly, scientists believe they might still be viable when humanity returns to the moon.

According to the theory of evolution, life diversified on Earth very slowly, with a single-cell creature developing into countless complex ones across eons of time. The ordering of fossils in Earth's rock layers—supposedly laid down over hundreds of millions of years—is said to demonstrate this process. But does the evidence actually support this claim?

The Grand Canyon is known worldwide as one of the premiere places to observe these supposed millions of years of geologic time. Each of its rock layers is said to represent a different period of evolutionary history. One of its most famous fossil-bearing layers has been called the Cambrian, and the layer just beneath that has been called the Precambrian.

Early geologists quickly discovered that the Cambrian rock layer, which they claim was formed 538 million years ago, contains an abundance of fossils with complex body structures, but the Precambrian directly beneath it has virtually none. Thus, passing from the Precambrian to the Cambrian

Comparing all of the supposed evolutionary time to one hour on a clock, the Cambrian explosion would cover less than one minute. The amount of time needed for evolution to accommodate an explosion of complex organisms in so many varieties is simply too long for any biological processes observed today.

seemed to show a sudden "explosion" of fossils. They named this phenomenon the "Cambrian explosion," as virtually all the modern divisions of animals first appear in the Cambrian layer.

When explaining his theory of slow evolution over time, Charles Darwin said, "All the organic beings which have ever lived on this earth may be descended from some one primordial form." He expected to find evidence of this gradual development of life arising from a common ancestor, linking to new life forms in a great evolutionary tree with hundreds of branches through the successive layers of Earth's rock. Yet rather than showing the slow evolution of simple life forms to complex ones, the fossils reveal the exact opposite. Famed Harvard evolutionary biologist Stephen Jay Gould would later admit, "The extreme rarity of transitional forms in the fossil record persists. ... The evolutionary trees that adorn our textbooks have data only at the tips and nodes of their branches; the rest is inference ... not the evidence of fossils."

Darwin also freely admitted that the actual evidence was a problem to his theory, writing, "To the question of why we do not find rich fossiliferous deposits belonging to these assumed earliest periods prior to the Cambrian system, I can give no satisfactory answer." The Cambrian explosion, which supposedly happened over a period of 10 to 15 million years, is too small a slice of time for the needed mutations to create the diverse animal life we see today.

Put simply, the number of sudden fossils appearing in the Cambrian layer falsifies gradual evolution—but it is no problem if you believe in a Master Designer. "For by Him all things were created that are in heaven and that are on earth, visible and invisible. ... All things were created through Him" (Colossians 1:16).

Not Enough Time

Another problem for evolution is the vast number of "missing links" that should be seen in rock layers. Significant groups of complex organisms appear suddenly without evidence that they morphed from something else. Regarding these, Darwin admitted, "Geology assuredly does not reveal any such finely graduated organic chain; and this, perhaps, is the most obvious and gravest objection which can be urged against my theory." Darwin believed that over time, geologists would begin finding the fossil evidence to prove his theory—but even after 150 years, such missing ancestors are still absent.

SOURCES

The Origin of Species, 6th edition, pp. 289, 291, 292

"Evolution's Erratic Pace." *Natural History*, vol. 86, May 1977, p. 14.

Earth's Lucky Atmosphere

An Amazing Fact:

Earth's atmosphere is a delicately balanced layer of air that helps make life possible on our planet. It is about 60 miles thick and comprises 78 percent nitrogen, 21 percent oxygen, and 1 percent other gases. This unique band of air also serves as a blanket insulating us from the sun's extreme heat and lethal radiation on one side—and the deadly cold of space on the other.

How did the Earth's amazing atmosphere form?

Opposing views exist on the origins of the mass of air that surrounds our world, in which we breathe, birds fly, clouds float, and sound travels.

> The term "firmament" comes from the Hebrew word *raqia*, which can also be translated as "expanse."

Evolutionists believe that billions of years ago, primordial organisms evolved into algae, which used the energy of our sun to produce oxygen. Over a few hundred million years, oxygen levels rose and destroyed some organisms while others adapted and thrived. This theory assumes a "reducing"—low—oxygen atmosphere since higher oxygen levels would have burned up early-evolving molecules.

However, the evidence reveals a different story. Rock samples found in some of the lowest sedimentary strata show the presence of oxygen—via oxidized metals in rock chemicals that often appear red (think of rusty iron). When geologists at Pennsylvania State University drilled into deep rocks searching for samples untouched by oxygen, they continued finding whole units of red and crystalized virgin rock that must have been formed in the presence of oxygen.

Is Earth's atmosphere really a product of billions of years of lucky evolution? Many features of our privileged planet's mass of air indicate a tuned and balanced design from the beginning—specifically allowing life to be sustained. The flawlessly complex arrangement of our atmosphere plays a vital role in dampening the harsh rays of the sun and moderating temperatures.

Additionally, along with its magnetic fields, the Earth's balanced atmosphere protects our planet from constant cosmic rays and charged solar particles that bombard the solar system in waves. Without this divinely created forcefield, all life on Earth would be destroyed by solar radiation.

The evidence in rocks and fossils does not support the notion of long periods of an oxygen-free atmosphere at any time on Earth—rather, the evidence is in harmony with the Bible. On the second day of creation, "God made the firmament, and divided the waters which were under the firmament from the waters which were above the firmament; and it was so" (Genesis 1:7).

Breathing Planet

Another dynamic way the atmosphere responds to changing conditions in space is through the winds. Water that evaporates from the ocean is picked up by air currents, absorbing and releasing heat and creating clouds that scatter sunlight. Atmospheric winds carry rain over large areas of land, preventing them from becoming lifeless deserts while also transferring heat to cooler areas. This "atmospheric circulation" contributes to the redistribution of thermal energy, making our planet habitable.

Symbiotic Relationships: A Miracle of Evolution?

An Amazing Fact:

Figs really are filled with wasps, but before you toss out your dried fruit, there are a few things to know! The female fig wasp is essential to a fig tree's reproductive cycle—she is the tree's only pollinator. In exchange for pollinating a fig, the wasp receives a safe nursery to hatch her young. After laying her eggs inside the fruit, she dies, and the fig's enzymes dissolve her body. Thus, the crunchy parts you chew in a fig are seeds—not parts of a wasp.

Yucca moth
(*Tegeticula yuccasella*)

This cooperation between fig plants and fig wasps is a form of **symbiosis** called "obligate mutualism," when both species depend upon the arrangement for their survival. Another example of this is the yucca plant and the yucca moth in the Southwest United States. Without the moth, the plant would lose its only pollinator; without the plant, the moth would lose its only food source.

Whether it's a relationship between flowers and pollinators, such as animals carrying seeds in their fur and dispersing them, or the living bacteria in your gut that help you digest food, we see mutualism everywhere. Think of the colossal exchange of plants taking in carbon dioxide and giving off oxygen and animals taking in oxygen and giving off carbon dioxide. Countless symbiotic relationships can be observed all over the world.

But how could so many cases of obligate mutualism evolve? If the moth needs the plant *and* the plant needs the moth, which one came first—and how did it survive without the other? The chances of this happening by accident are so close to zero that most statisticians would consider it impossible. Evolution simply cannot explain the many mutualistic relationships found in our world!

Aren't such relationships best explained as the mark of a skillful designer? Indeed, it is found in the very character of our world's Creator! Jesus once said, "Give, and it will be given to you: good measure, pressed down, shaken together, and running over will be put into your bosom. For with the same measure that you use, it will be measured back to you" (Luke 6:38). And the apostle Paul wrote, "Bear one another's burdens, and so fulfill the law of Christ" (Galatians 6:2). Cooperative giving and receiving is a heavenly principle! "For since the creation of the world His invisible attributes are clearly seen, being understood by the things that are made, even His eternal power" (Romans 1:20).

Clownfish and sea anemones share a symbiotic relationship. The fish protects the anemone from predators—and, in turn, the anemone's stinging tentacles protect the clownfish from its predators. The clownfish is coated with a special mucus that protects it from the stinging tentacles.

Macroevolution vs. Microevolution

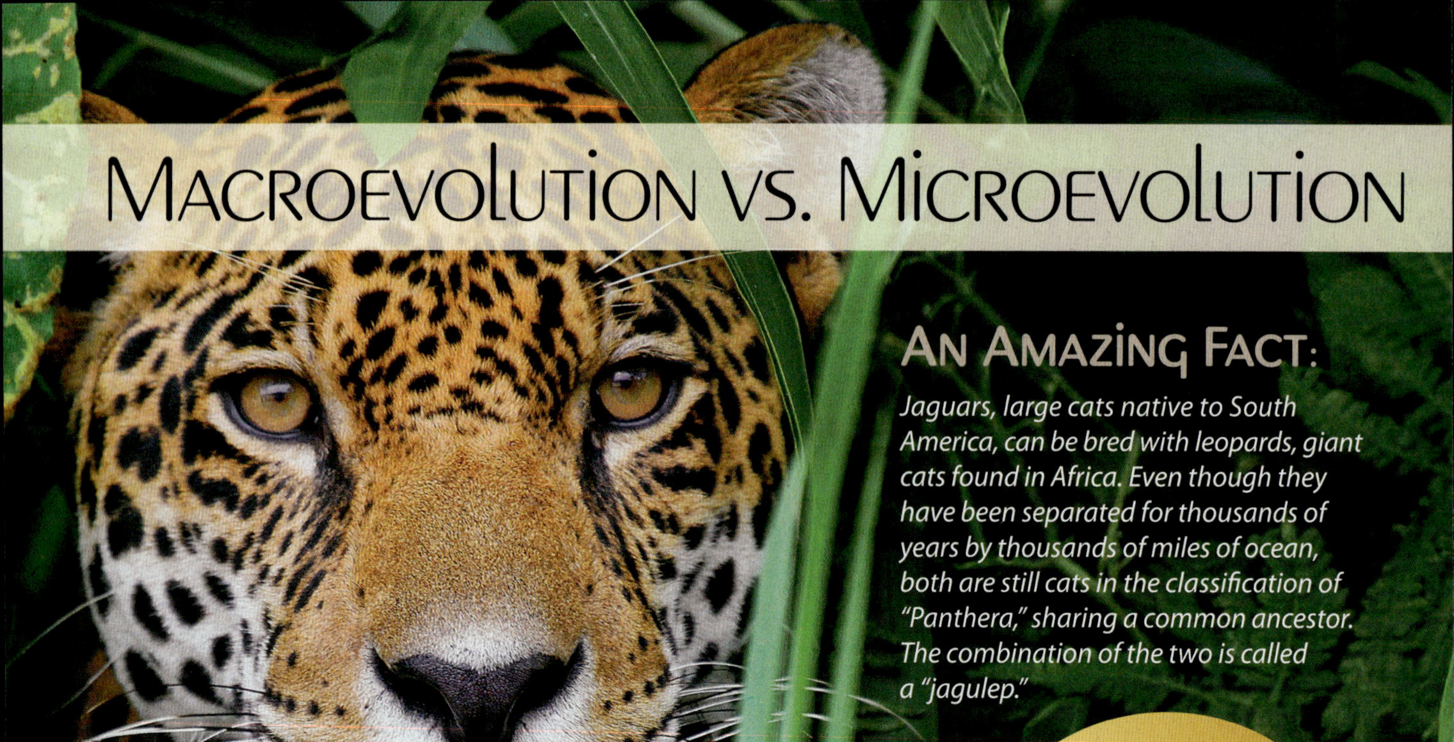

An Amazing Fact:

Jaguars, large cats native to South America, can be bred with leopards, giant cats found in Africa. Even though they have been separated for thousands of years by thousands of miles of ocean, both are still cats in the classification of "Panthera," sharing a common ancestor. The combination of the two is called a "jagulep."

Photo courtesy of Shy Wolf Sanctuary Naples, Florida

"Bango," a jagulep hybrid.

© Carolyn Collmer

Many are surprised to learn that most Creationists believe in evolution. It's true!

But hold on—when you ask Creationists if they believe in "evolution," you'll likely get this response: "It depends on what you mean by 'evolution.'" Well, what exactly do they mean?

Simply put, evolution—that is, change over time—can be divided into two categories:

Macroevolution is the belief that a species can transform into an entirely new species over time, a process called "speciation." This happens in stages and is said to be the force that turned an amoeba into human beings over billions of years. When most people hear the word "evolution," they think of it in this way.

The other category is **microevolution,** the idea that small adaptive changes occur within species due to pressures in their natural environment. These changes can happen fairly quickly, sometimes in a single generation. The word "adaptation" is a synonym for microevolution.

An example of macroevolution would be a reptile turning into a bird over millions of years—a total leap between species. An example of microevolution would be when the fur of an Arctic hare turns white in the winter to help it camouflage better in the snow; its ears also become shorter to conserve body heat. Its cousin down south, by comparison, the desert hare, features brown fur to camouflage it in the desert and longer ears to help it cool more efficiently.

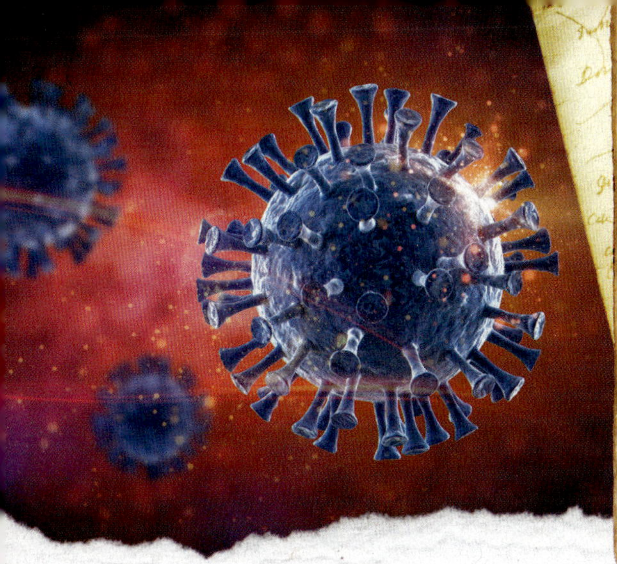

Some say that Creationists deny "natural selection" when they deny the theory of evolution. But it's not true—because Creation scientists observe changes within species every single day! Think of a mutating virus, like COVID-19, which rapidly changes to survive and spread to more hosts. So, what's the problem? Creationists know that science has never documented a virus evolving into something other than a virus. No matter how many strains of COVID-19 "evolved" every week, it was still a virus.

Microevolution is the only kind of evolution to have been scientifically observed. While studying birds in the Galapagos Islands, Darwin noted many differences and variations in the beaks of finches. This was microevolution; they all stayed finches.

But no one has ever witnessed macroevolution—that is, no one has observed one species changing into another. They once hypothesized that enough instances of natural selection would eventually lead to macroevolution. Macroevolution relies heavily on "forensic" science—piecing together fragments of bones and declaring that an entirely new species arose. However, it's based on conjecture and belief, making it pure fantasy!

You can see significant changes in cats and dogs as part of selective breeding—but the diminutive Chihuahua and the giant Saint Bernard are both dogs. Over the centuries, the Dingo in Australia evolved to survive in its environment more efficiently; likewise, over the same time, the Siberian Husky evolved with a longer coat to adapt to its environment. No matter whether a breeder "evolves" a dog from a rottweiler to a poodle, they remain the same species. These are all cases of microevolution.

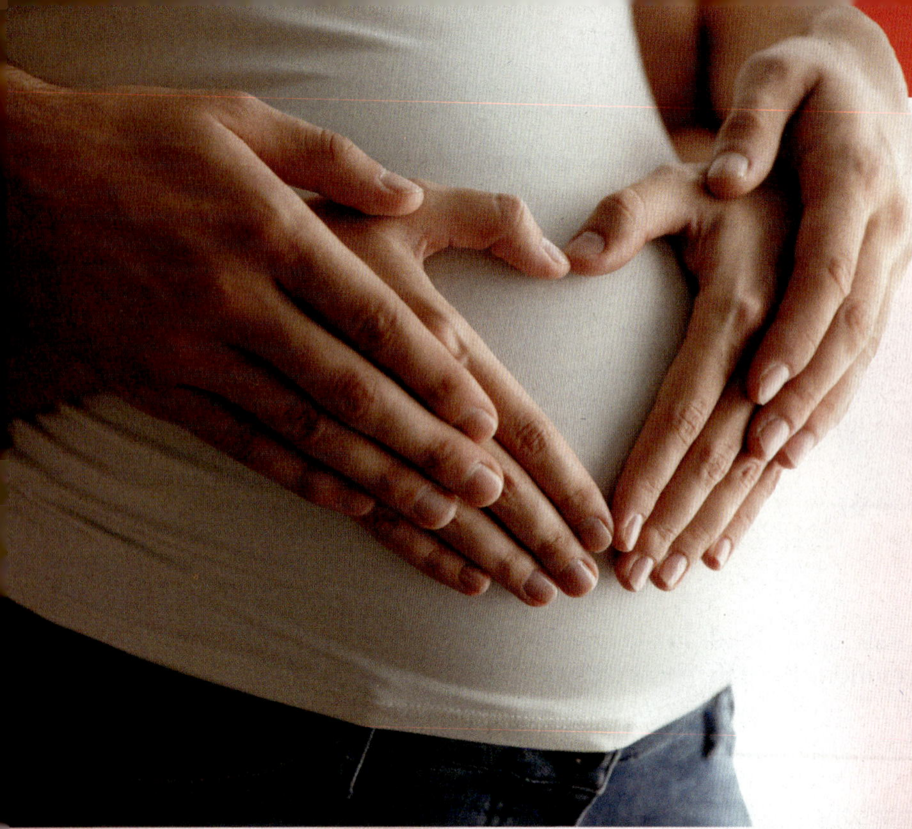

An Amazing Fact:

During the first month of pregnancy, a human zygote transforms at a dazzling pace: from a single cell to a complex organism with a beating heart and developing nervous system. By the eighth week, it's termed a fetus, with all major organs formed. Remarkably, this intricate process hinges on precise genetic instructions, orchestrating the emergence of more than 200 different cell types that ultimately coalesce into a fully formed baby in just nine months.

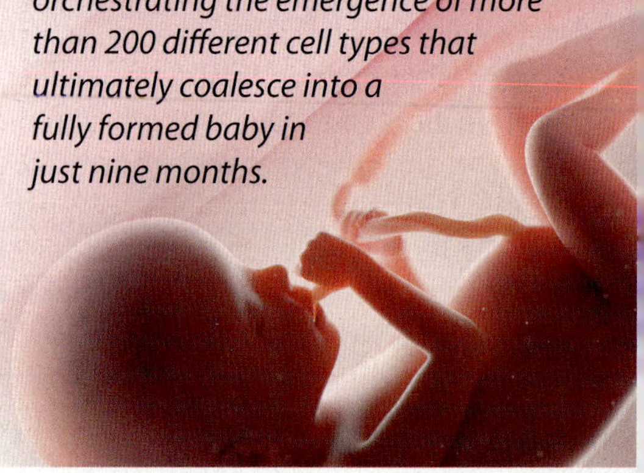

Sexual Reproduction:
A Mystery of Biology

Sex has been called one of the great mysteries in biology and a major unsolved problem. Why? The "adaptive advantage" of sexual reproduction does not fit evolutionary theory. Having two sexes, male and female, for many species is not advantageous to multiplying. To be asexual makes much more sense from a survival-of-the-fittest perspective. It's far more efficient and requires only one organism.

Sexual reproduction involves significant costs. It requires finding a mate, which can be time-consuming and risky, as sexual displays to attract a mate can put an organism at risk of being killed. It cuts the field in half for potential mates, and only half of an organism's offspring (typically females) can reproduce, slowing population growth. Additionally, recombination during sexual reproduction can disrupt advantageous gene combinations.

And what do we make of the pleasure of intimate acts and relationships? How did that evolve?

Different theories have been offered to explain sexual reproduction from an evolutionary standpoint, but a purely "natural" (random) process cannot adequately explain the development of the two sexes. Those who emphasize a "need" for two sexes (to mask, for instance, mutational damage of the genome) point to a process that is intelligently guided, which, ironically, implies an intelligent designer.

Even more challenging is trying to explain the evolution from asexual organisms to sexual organisms and the significant changes that would need to take place in the DNA. Then, there is the question of why we still have asexual organisms. Evolutionist Graham Bell admitted that "sex is the queen of problems in evolutionary biology."

One fascinating observation that implies the involvement of a divine influence in human reproduction is that almost the same number of boys as girls survive to marriageable age. Married couples have a slightly higher chance of having a baby boy than a girl. According to the Pew Research Center,

SOURCES

Graham Bell, *The Masterpiece of Nature: The Evolution and Genetics of Sexuality* (Berkeley: Univ. of California, 1982), 19.

https://www.pewresearch.org/religion/2022/08/23/sidebar-sex-ratios-around-the-world.

there are approximately 105 male births per 100 female births worldwide. This is largely because more females die in miscarriages. However, because males tend to take more life-endangering risks, especially in their younger years, the ratio is nearly one-to-one by the time both sexes reach marriageable age. Science cannot explain how random evolution could account for this ideal ratio.

Furthermore, the belief in evolution degrades romantic love into base physiological responses to chemical cues for the instinctual purpose of perpetuating the species. But Solomon described romantic love as a most wonderful joy: "There are three things which are too wonderful for me, yes, four which I do not understand: the way of an eagle in the air, the way of a serpent on a rock, the way of a ship in the midst of the sea, and the way of a man with a virgin" (Proverbs 30:18, 19). It appears that we were created for companionship!

The Bible teaches that in the beginning, at Creation, God made humans "male and female" (Genesis 1:27). The Lord also made plants and animals to reproduce "according to [their] kind" (vv. 11, 12, 21, 24, 25). That means both sexual and asexual organisms were part of God's original plan. Rather than trying to explain the mystery of sexual reproduction through confusing and contradictory hypotheses, the Bible's explanation makes the most sense.

In the field of evolutionary biology, the existence of sexual reproduction in the natural world is sometimes called the "paradox of sex." This is because sexual reproduction can dilute genetic representation in offspring. To evolutionary biologists, it's unclear why sexual reproduction exists and persists despite its apparent disadvantages compared to asexual reproduction.

Irreducible Complexity

An Amazing Fact:

Coral reefs provide habitat, food, and breeding grounds for about 25 percent of all marine species. They are among the planet's most delicately balanced and diverse ecosystems, containing thousands of plant and animal types. A vital component of coral reefs is soft-bodied organisms known as coral polyps. The removal or death of this single organism leads to the breakdown of the reef structure, impacting countless species.

One pathway for supporting the existence of God is to observe the natural world and see just why living things are best explained in the context of an intelligent designer. Even the Bible says, "Since the creation of the world [God's] invisible attributes are clearly seen, *being understood by the things that are made*" (Romans 1:20, emphasis added). Indeed, while most scientists label anything that suggests an intelligent deity as "false science," many other scientists believe that the natural world's evidence supports intelligent design.

A significant reason for this belief is the concept known as **irreducible complexity**, which states that there are hundreds of biological systems—plants and animals—with many interacting and sophisticated parts. If even one of these parts were removed or reduced, the system could not survive or exist. Statistically, it's nearly impossible that any of these systems evolved through small, step-by-step modifications of earlier, less-complex systems through natural selection.

For example, the eggbeater, a common kitchen tool, is made

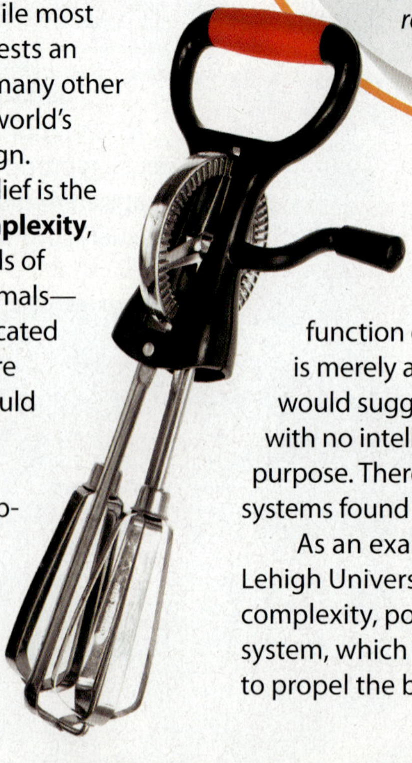

up of four parts: a handle, a turning knob, rotating gears, and whisks. If any one of these parts is missing, the function of the eggbeater ceases. An eggbeater is merely a simple mechanical device, but nobody would suggest its parts came together by accident with no intelligence assembling them for a specific purpose. There are much more complex mechanical systems found in nature.

As an example, Professor Michael Behe of Lehigh University, who coined the term irreducible complexity, points to the E. coli bacteria's flagellar system, which has 40 individual, integral parts used to propel the bacteria from one place to another.

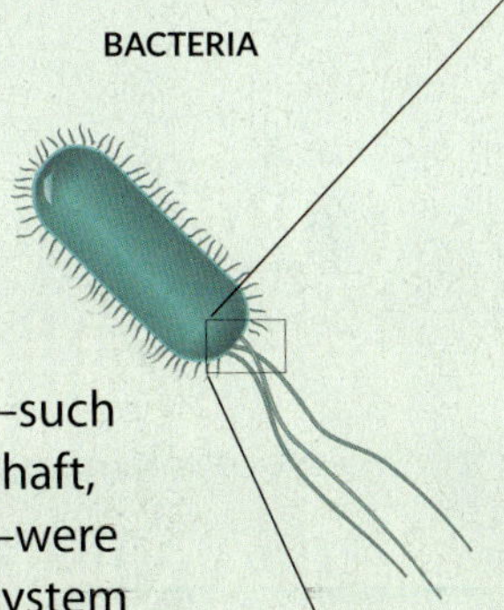

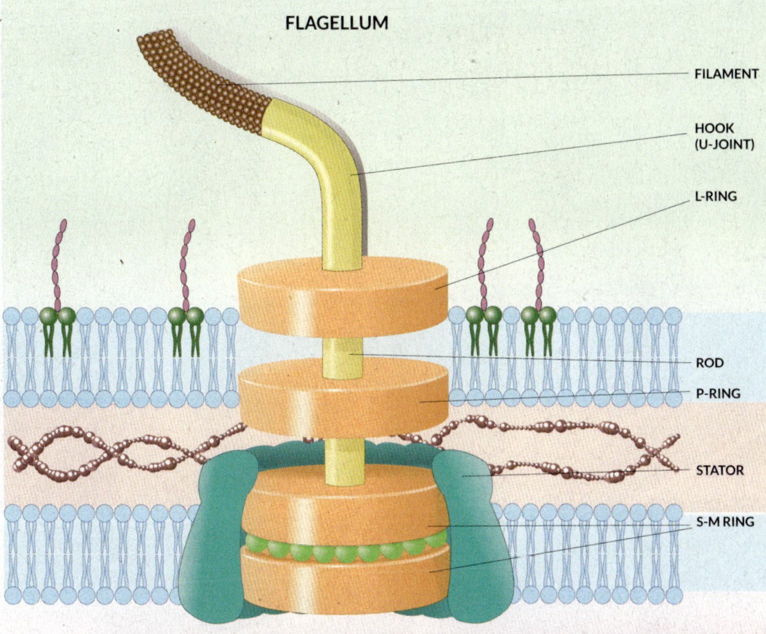

If any of these parts—such as stator, rotor, driveshaft, u-joint, or propeller—were removed, the entire system would fail to function. Charles Darwin even admitted,

> If it could be demonstrated that any complex organ existed, which could not possibly have been formed by numerous, successive, slight modifications, my theory would absolutely break down (*Origin of Species*, 1859, p. 158).

Irreducible complexity reveals that some biological systems are so intricate that it's impossible for them to have evolved independently by chance. The system would be incomplete and useless unless all parts fully evolved simultaneously. According to the laws of evolution, such an incomplete system would be naturally selected out of the organism. Even the DNA replication and repair that happens in every living animal is accomplished by molecular machines that could never evolve independently; this points to the involvement of an intelligent designer with a higher purpose.

Something more than a mere random process is involved in the origin of these highly sophisticated organic machines that propel bacteria, repair DNA, and enable life on Earth. The Bible says of this stunning reality, "Such knowledge is too wonderful for me; it is high, I cannot attain it" (Psalm 139:6).

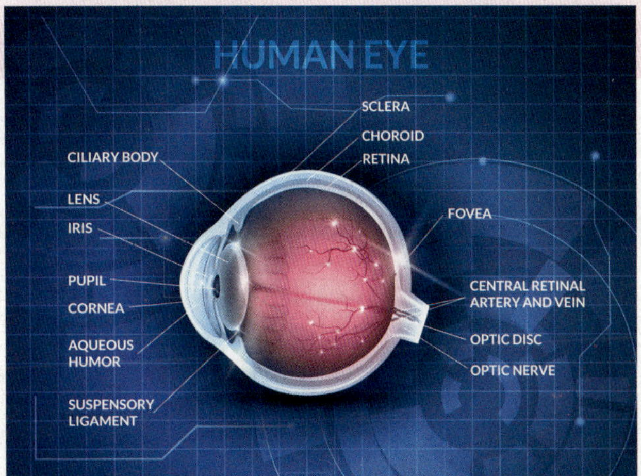

The human eye is a prime example of irreducible complexity. Its interconnected parts include the lens, retina, optic nerve, cornea, iris, and vitreous humor. The eye could not have evolved through incremental steps because it does not function unless all its parts are intact and working together.

The blood clotting system, also known as the coagulation cascade, aptly demonstrates irreducible complexity. This cascade includes multiple steps and components such as platelets, clotting factors, fibrin, and regulatory proteins. The many steps involved in blood clotting are so complex that they can only function if all the components are working at the same time.

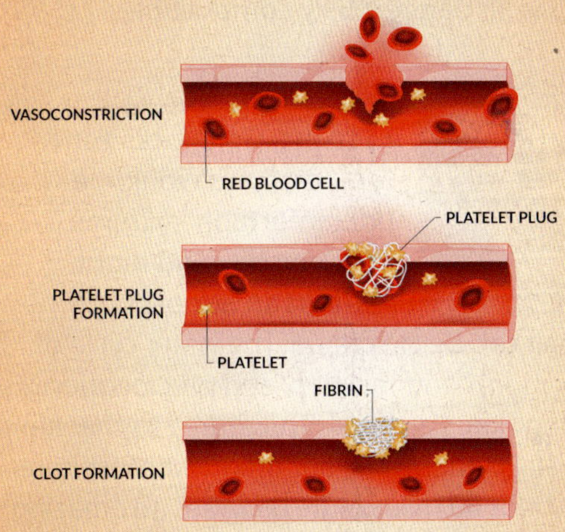

Our Goldilocks Planet: Just Right

An Amazing Fact:

Earth orbits the sun at a distance of nearly 94.5 million miles, traveling at about 67,000 mph and spinning at about 1,000 mph. These conditions are perfect for sustaining life on our planet. Any significant changes to them would have deadly catastrophic effects.

No doubt, you have heard the fairytale story of Goldilocks, the little girl who intruded into the home of three bears while they were out taking a walk to let their porridge cool. While the bears were gone, Goldilocks gobbled up the baby bear's porridge, which was not too hot or too cold—but just right. She then sat in the baby bear's chair, which was not too big or too hard—but just right. And then she fell asleep in the bed that wasn't too high or too low—but just right.

The **Goldilocks zone** is a term in astrobiology that refers to the habitable area in relation to a star. Earth's distance from the sun just happens to be in the perfect habitation zone—not "too hot" (like Venus), not "too cold" (like Mars)—but "just right."

Evolutionists struggle to explain how Earth became a Goldilocks planet. How improbable is it that such a planet came to exist in a relatively tiny habitable zone—and then stayed there for a few billion years to allow life to evolve? It all happened by "luck," "chance," or "accident," they say.

But a habitable temperature zone—one that allows liquid water to exist, neither boiling away nor freezing solid—is just one of many ideal things needed to sustain life. Let's consider how Earth's atmosphere, sun, and moon are situated "just right."

ATMOSPHERE

Earth's atmosphere has just the right combination of gases—about 78 percent nitrogen and 21 percent oxygen. If the nitrogen concentration were higher, there would be too little oxygen to sustain the life of respirating creatures. On the other hand, a concentration of too much oxygen would make the air dangerously combustible.

TEMPERATURE

Even in a Goldilocks zone, temperatures would be uncomfortably extreme between day and night without an atmosphere where air can be circulated. The water cycle of evaporation, condensation, and precipitation also helps regulate Earth's temperatures so that life can exist. Furthermore, our unique atmosphere, rising about 60 miles high, provides an amazing shield from the sun's blistering radiation.

Yes, Even the Lightning

Earth also has an ideal balance of atmosphere, temperature, and moisture that creates an environment where there are about 6,000 lightning strikes every minute, which is more than 8 million strikes every day. Thunderstorms and electrified clouds are like batteries that cause the earth to have a negative charge and the atmosphere to have a positive charge. Thunderstorms help transfer the negative charges back to Earth. Without thunderstorms and lightning, the electrical balance in Earth's atmosphere would disappear in minutes. Lightning also makes helpful ozone-producing chemicals.

PLANET SIZE

Then there is the perfect size of our planet, with an iron and nickel core that gives us just the right amount of gravity. If Earth were too big, gravity would crush us; if it were too small, we would float off into space.

STAR SIZE

Some astrobiologists believe that if the sun were a more massive star, it would emit too much ultraviolet radiation for organic molecules to survive, even if our planet were positioned farther out to sustain habitable temperatures. On the other hand, if the sun were smaller, its habitable temperature zone would be closer in, creating another problem: gravitational lock (tidal lock). A stronger gravitational pull would cause our planet to spin too slowly, or not at all, keeping one side burning and the other freezing. Life could survive only along a thin longitude between light and darkness. It would be like Mercury, which rotates once every 59 Earth days, or Venus, which rotates once every 243 Earth days.

MOON SIZE

The size of the moon and its distance from Earth is "just right" to control oceanic tides. Each day has two high tides and two low tides, following approximately a 25-hour cycle. If the moon were smaller or farther away, the tides would be too weak to sufficiently circulate and oxygenate the coastal waters in bays and estuaries, and the oceans would become stagnant. On the other hand, if the moon were larger or closer, the tides would be too powerful, overwhelming the coasts with destruction. Also, consider how incredible it is that because the moon is 400 times smaller than the sun, yet 400 times closer to Earth, the lunar and solar discs appear the same size —an optical illusion described by Isaac Asimov, a science fiction writer, as "the most unlikely coincidence imaginable." Earth is the only known planet in which a total solar eclipse enables us to study the sun in a way that would otherwise be impossible.

Did all this happen by chance?

A more probable scenario is that a super-intelligent being designed Earth's atmosphere, cycle, position, the sun, and the moon in a way that sustains life. In the Genesis account of a six-day creation, we read that God created a firmament, or expanse, "and divided the waters which were under the firmament from the waters which were above the firmament" (Genesis 1:7). Then God said, "Let there be lights [sun, moon, and stars] in the firmament of the heavens to divide the day from the night; and let them be for signs and seasons, and for days and years" (v. 14). The result of all this was the perfect pocket of atmosphere to sustain life, and this was made so that humans, the finishing touch of His handiwork (vv. 26–31), could thrive.

Mitochondrial Eve: The First Mother

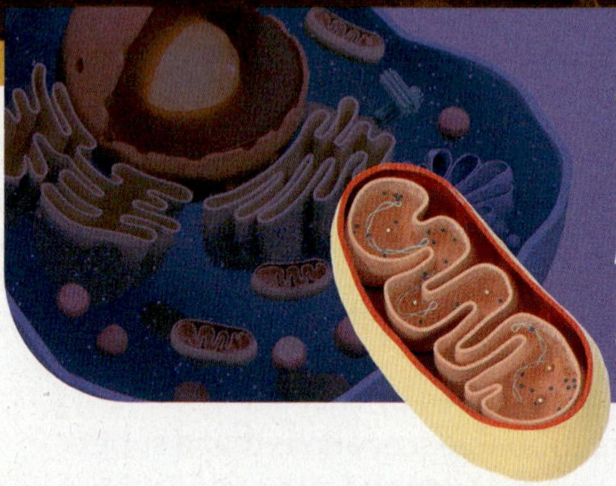

An Amazing Fact:

Mitochondria are the "batteries" found in every cell of our bodies. They provide energy to keep us working properly. Your body contains about 25 thousand trillion mitochondria! They are so small that you can fit 2,000 of them on the head of a pin.

Can humans today trace their lineage back to one common ancestor, as the Genesis account suggests, to one mother of all humans? Evolutionary scientists now believe that all living humans can be linked genetically to a single maternal "Eve" who lived about 200,000 years ago.

She is believed to be the most recent woman from whom all living humans descend in an unbroken line purely through their mothers and through the mothers of those mothers, back until all lines converge on this one woman.

Geneticists have studied a small and unique chromosome inside the mitochondria, a tiny organelle that provides energy to a human cell. A loop of DNA passed from mother to child over generations provides information that could be used to peek into humanity's past.

However, ongoing research has revealed that data alone cannot determine how long ago the first mother of all humanity lived. Starting "assumptions" are necessary to calculate time frames. The age of **Mitochondrial Eve** will vary depending on which assumptions are used. Some assumptions for these genetic clocks calculate that our first ancestral mother lived 100,000 to 200,000 years ago, while others state she lived only 6,000 years ago. One research project openly states that "how

World Population Milestones Throughout History

YEAR	MILESTONE	YEARS SINCE PREVIOUS MILESTONE
1804	1 billion	All of human history
1927	2 billion	123
1960	3 billion	33
1974	4 billion	14
1987	5 billion	13
1998	6 billion	11
2010	7 billion	12
2022	8 billion	12
2037 *(projected)*	9 billion	15 *(projected)*

they [humans] evolved in time depends upon the model of evolution that you use." Exactly!

If we honestly consider the facts, it's inconceivable that humans have been wandering the planet for 100,000 years without much increase in population—and then suddenly, the population exploded from about a billion in 1804 to 8.1 billion in just 220 years.

The Bible says, "Adam called his wife's name Eve, because she was the mother of all living" (Genesis 3:20). If one considers the abundant evidence in nature that supports a young Earth and the record of biblical genealogies, it is reasonable to believe that life on Earth began approximately 6,000 years ago. And if we believe in the Bible account, we are not left to guess where we came from.

SOURCES

Rice University. "'Mitochondrial Eve': Mother of all humans lived 200,000 years ago." *ScienceDaily*, 17 August 2010

https://en.wikipedia.org/wiki/Mitochondrial_Eve

https://populationconnection.org/learn/population-milestones/

The Flood

Earth's Population

An Amazing Fact:

In 1859, Thomas Austin, a wealthy settler who lived in Victoria, Australia, imported 24 European wild rabbits, which he let roam free on his estate for occasional hunting. The rabbits quickly ventured from his backyard sanctuary, however, and began to multiply. It took only a decade for these cute but invasive (meaning non-native) creatures to become 20 million. Now, over 200 million rabbits are spread across 70 percent of Australia's landmass.

Did you know that the current population of the world can help us determine the approximate age of the earth? Here's how …

There are about 8 billion people on the planet today. Because the human population doubles globally approximately every 150 years—a conservative estimate that accounts for plagues, wars, and natural disasters—we can calculate back through time to get an estimate of how long humanity has been around.

Interestingly, today's number makes the best sense if the earth is only about 6,000 years old. (Remember, there was a catastrophic flood around 2500 BC, but 4,500 years is enough to account for our current population.) But if humans have existed for 50,000 or more years, multiple trillions of people would be spilling off the planet!

Bible Numbers

Consider the history of Israel. God promised the children of Abraham that if they would obey Him, He would help them multiply rapidly:

> He will love you and bless you and multiply you. … You shall be blessed above all peoples; there shall not be a male or female barren among you. … And the Lord will take away from you all sickness, and will afflict you with none of the terrible diseases of Egypt which you have known (Deuteronomy 7:13–15).

According to Genesis 46:27, when Jacob and his family moved to Egypt, they numbered 70 people. But within 215 years (from Jacob's entry into Egypt till the Exodus, according to various Bible scholars), their

numbers exploded, which explains why the king of Egypt was terrified.

> Now there arose a new king over Egypt, who did not know Joseph. And he said to his people, "Look, the people of the children of Israel are more and mightier than we; come, let us deal shrewdly with them, lest they multiply, and it happen, in the event of war, that they also join our enemies and fight against us, and so go up out of the land" (Exodus 1:8–10).

Despite Pharaoh's efforts to stifle their growth, "the children of Israel were fruitful and increased abundantly, multiplied and grew exceedingly mighty; and the land was filled with them" (v. 7). After the Hebrews left Egypt, Moses took a census of only the fighting men, and they numbered just over 600,000 (Numbers 1:46). (That number may include the "mixed multitude," non-Israelites who were part of the Exodus.) Adding the women and children, the number of people may have been between 1.8 and 2.4 million! This gives us a rough idea of how quickly people can multiply.

The population of the United States in 1776 was 2.5 million people. Now, only 250 years later, it is 341 million.

Starting Over After the Flood

> Behold, I Myself am bringing floodwaters on the earth, to destroy from under heaven all flesh in which is the breath of life; everything that is on the earth shall die (Genesis 6:17).

No other catastrophe in the Bible is quite like the one found in Genesis chapters 6–8, the account of Noah's ark and a worldwide flood—an event that destroyed all land animals except for those preserved on the ark. The Bible explains:

> Then the Lord saw that the wickedness of man was great in the earth, and that every intent of the thoughts of his heart was only evil continually. And the Lord was sorry that He had made man on the earth, and He was grieved in His heart. So the Lord said, "I will destroy man whom I have created from the face of the earth, both man and beast, creeping thing and birds of the air, for I am sorry that I have made them" (Genesis 6:5–7).

The furious nature of this worldwide flood surely changed the face of the earth, so wouldn't we see proof of this devastation on the earth today? Later, we will consider evidence from science of the plausibility of a worldwide flood.

The Lord called Noah "a just man, perfect in his generations" (v. 9). He asked this faithful servant to build an ark, an enormous boat built for animals, food, and people. Then God said, "I will cause it to rain on the earth forty days and forty nights, and I will destroy from the face of the earth all living things that I have made" (Genesis 7:4).

After no other humans responded to Noah's pleas, he, his wife, and their three sons and their wives entered the ark. After seven days, "all the fountains of the great deep were broken up, and the windows of heaven were opened" (v. 11). Describing this global deluge, the Bible says,

> The waters prevailed exceedingly on the earth, and all the high hills under the whole heaven were covered. The waters prevailed fifteen cubits upward, and the mountains were covered. And all flesh died that moved on the earth: birds and cattle and beasts and every creeping thing … and every man. … So He destroyed all living things which were on the face of the ground: both man and cattle, creeping thing and bird of the air. They were destroyed from the earth. Only Noah and those who were with him in the ark remained alive (vv. 19–21, 23).

Is it reasonable to believe an event of this proportion could have occurred? Some suggest that this story describes only a local flood, but the text plainly indicates a violent deluge of drastic proportions, with waters covering mountains, killing every living thing on the earth. If it was a local flood, why would Noah be instructed to build a massive boat? Wouldn't he simply need to move his family to some other place? And why would God later make a covenant through the rainbow that "the waters shall never again become a flood to destroy all flesh"? (Genesis 9:15).

From the population trends observed throughout history and what we see on the earth today, we can conclude that humanity's population is best understood when going back about 4,500 years, which is when the Bible tells us that humans began populating the planet.

Today's Population and the Flood

Regarding the incredibly small human population growth in terms of evolution theory, scientists admit, "You've got to come up with a hypothesis for an event that wiped out the vast majority of that variation, … something [that] caused the human population to drop drastically. When or how often that may have happened is anybody's guess. Possible culprits include disease, **environmental disaster,** and conflict" (Lee Dye, "We Dodged Extinction." *Dye Hard Science.* ABCnews.com, May 26, 1999).

Soft Tissue in Old Bones

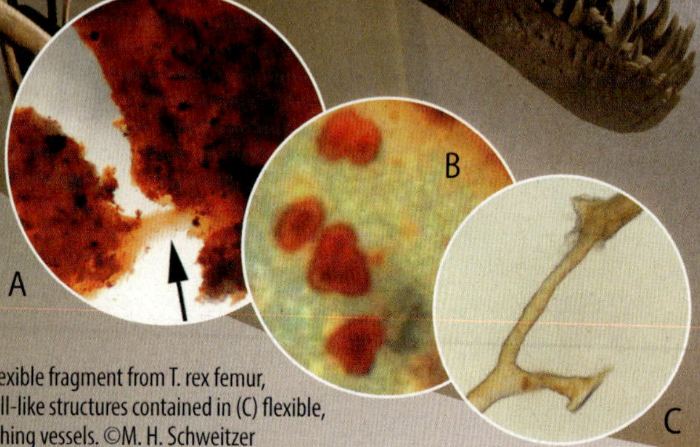

(A) Flexible fragment from T. rex femur, (B) cell-like structures contained in (C) flexible, branching vessels. ©M. H. Schweitzer

An Amazing Fact:

In 2005, a team of scientists led by Dr. Mary Schweitzer, a paleontologist at North Carolina State University, discovered soft tissue—that is, pliable blood vessels—in the bones of a Tyrannosaurus rex. The shocking findings were published in the journal Science *and shook the evolutionary community to its core.*

The recent discoveries of soft tissue in fossils that are reportedly millions of years old have seriously challenged evolutionary assumptions about the age of the earth and how fossils are formed. Indeed, this evidence supports the view that the earth is younger than evolutionists claim—much younger than their fossil record dating.

Not only was soft tissue recovered in 2005 from a Tyrannosaurus rex thigh bone, but in 2012, soft tissue was also found in the horn and rib of a Triceratops that was uncovered in Montana's Hell Creek Formation. These findings contradicted what evolutionary scientists had always believed about the preservation of soft tissue: Proteins that make up soft tissue should degrade after one million years. Yet the fossils containing that tissue are dated to be more than 60 million years old!

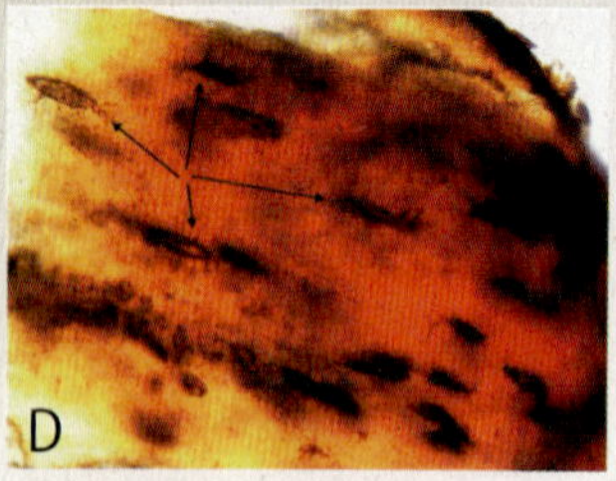

Portion of soft, stretchy fibrillar bone from Triceratops horn. ©Mark H. Armitage

How was this contradiction explained? In 2013, researchers concluded that iron from hemoglobin in red blood cells became highly reactive once the creature died, creating amino-acid "crosslinks" that made the proteins resistant to decay for millions of years. This idea was supported by a lab experiment in which ostrich blood vessels were preserved for two years in an iron-rich solution. However, the amount of iron needed to preserve soft tissue in this manner isn't naturally available.

In addition, the presence of soft tissue in dinosaur bones further challenges the reliability of radiometric dating. If such tissue is not as old as evolutionists claim it should be, then the dating methods used to determine the bones to be millions of years old can't be reliable either. Such is the case of the Tyrannosaurus rex bones, dated to be 68 million years old.

"The Right Conditions" Not a Magic Wand

Researchers now say that the presence of such tissue may not be as uncommon as previously thought. In a study published in *Nature* in 2017, collagen, a protein found in bone and connective tissue, was found in a dinosaur skeleton dated 195 million years old. Some argue this suggests that some forms of organic material can survive for millions of years under the right conditions—but even a fresh piece of bread left unwrapped will become brittle overnight.

SOURCES

https://www.cambridge.org/core/journals/microscopy-today/article/preservation-of-triceratops-horridus-tissue-cells-from-the-hell-creek-formation-mt/11CD094ED1312B6C618E098C12FCC324

Figures A, B, and C: *Science,* March 25, 2005. Reprinted with permission from AAAS.

Figure D: *Microscopy Today,* Jan. 8, 2016. Reprinted with permission from Oxford University Press.

Living Fossils

An Amazing Fact:

Regularly found as a fossil specimen, the coelacanth was long thought to have been extinct for more than 66 million years—then, in 1938, a live specimen was found by a fisherman off the coast of South Africa. Incredibly, the living version was identical to the fossils. The surprising discovery has inspired modern artwork and literature. Over 20 countries have featured this lobe-finned fish on postage stamps.

"Living fossil" is a term coined by Charles Darwin to describe organisms found in the fossil record that are still alive today, appearing to have changed little over a long period of time. The six-foot-long coelacanth, for example, with its dull coloration and white spots, is considered a living fossil because it looks just like its fossilized ancestors but does not have any close living relatives by evolutionary reckoning.

This unusual deep-sea fish, with armored scales and a hinged skull, first appeared, according to evolutionists, over 300 million years ago—70 million years before the first dinosaurs. Until the discovery of living coelacanths in 1938, it was assumed they were wiped out along with the dinosaurs.

More recently, scientists have claimed that modern coelacanths have acquired new genes in the last 23 million years, so they are not quite the same fish. But comparing the genome of living coelacanths with their dead fossil counterparts is impossible—the DNA is too degraded. Other explanations

Based on coelacanth fossils, scientists once thought that the four fins on the fish's underside would eventually evolve into legs. But when living ones were studied, those fins aided swimming and had not evolved for walking along the ocean floor.

based on comparisons of similar genes are merely guesswork.

More Living Fossils

Another fascinating "living fossil" is the horseshoe crab. Not a real crab, this unusual sea animal is more closely related to an arachnid, an eight-legged creature such as a spider or scorpion. Evolutionists claim the fossils of these crabs are up to 445 million years old. But how could these arthropods remain virtually unchanged for eons when everything around them experienced radical evolution?

What's fascinating about this "primitive" creature is its advanced set of eyes—ten in all. Some are used to find mates, some to detect ultraviolet light, and some to orient the creature when swimming. Furthermore, the horseshoe crab's unique blue blood is used in medicine to detect endotoxin, a component in some bacteria that can cause a severe immune response.

Then there are the amazing dragonfly fossils said to be 270 to 300 million years old—despite the abundance of living specimens in the world today. Moreover, these sophisticated flying creatures have nearly 360-degree vision and perfect aerodynamic design. In fact, the fossils depict them as once the largest flying insects ever, with a wingspan of up to 30 inches. If evolution is true, why have they remained basically the same?

Living fossils powerfully challenge evolutionary theory, which teaches that change will always occur over long periods of time. Yet we find many modern examples, besides dragonflies and horseshoe crabs, that have remained virtually unchanged for "millions" of years. Such creatures are living proof that macro-evolutionary change across species is a myth.

However, finding creatures that are effectively unaltered—allowing for *micro* changes *within* species as they adapt to their environments—makes sense to those who believe the Genesis account: "In the beginning, God created … the living creature according to its kind" (Genesis 1:1, 24).

Photo by Chris McGrath/Getty Images

The Wollemi pine, seen in fossils believed to date back 145 million years, was thought to have gone extinct with the dinosaurs until it was rediscovered by a group of hikers in 1994 in a national park west of Sydney, Australia. Less than 100 of these adult trees are known to exist in the wild.

What Happened to the Dinosaurs?

An Amazing Fact:

Scientists estimate that 100 to 10,000 species—from microscopic organisms to large plants and animals—go extinct yearly.

What caused the mass extinction of dinosaurs? The fossil record contains plenty of evidence of their demise. Jumbled dinosaur graveyards trapped in hardened silt have been unearthed and include bones, teeth, tracks, eggs, embryos, skin impressions, and even dung. According to evolutionary theory, dinosaurs existed on a global scale between 65 and 225 million years ago.

Paleontologists have identified nearly a thousand species of dinosaurs, and their fossils are found on every continent, including Antarctica. So, what killed these once-plentiful creatures? The theories have evolved.

Paleontologists once said there was a serious break in the food chain and that dinosaurs starved. Some claimed dinosaurs evolved into birds. Still, others thought there may have been a killer pandemic that spread through the giant reptile population.

If there were any dinosaurs preserved on the ark, they went extinct over time, likely due to the pressures of hunting and harsh environmental factors that occurred immediately after the Flood ended.

Even though most of the dinosaur bones discovered indicate that the animals died rapidly and were covered by mud, scientists flat-out dismissed the possibility of Noah's Flood being a culprit. However, because the evidence for the biblical event is so compelling, they came up with another theory—a giant asteroid hit the planet, triggering massive tsunamis (i.e., major flooding) and filling the skies with ash that blocked sunlight and cooled the planet.

It's worth mentioning that some creation scientists believe God may have used asteroids to crack the earth's crust, triggering the flooding, and "all the fountains of the great deep were broken up" (Genesis 7:11). The 56-mile-wide Acraman impact crater in South Australia is one example of evidence supporting this idea.

The theory of evolution states that dinosaurs evolved from archosaurs through a gradual process of mutations and natural selection. This means we should find more complex organisms above a particular rock layer in the geologic column and less complex ones below. But the fossil record doesn't support this. Instead, dinosaur fossils appear suddenly in the rock layers without any connection to previous ancestors and are fully functional to live in their environment.

The global nature of dinosaur extinction makes it clear that the world's climate used to be much more uniformly mild and that a global catastrophe best explains what happened to these creatures. The deluge described in Genesis 7 tells that the extinction of animals outside the ark happened quickly and not over millions of years. The scientific evidence points to a short chronology of Earth's history that agrees with this Bible record.

A Flood of Legends

An Amazing Fact:

Why does Michelangelo's marble sculpture of Moses, created for the tomb of Pope Julius II in Rome, show the ancient Bible character with horns protruding from his head? When St. Jerome translated the Bible into Latin in the late 4th century, he read how Moses' face shone when he descended from the mountain with the stone tablets (Exodus 34:29, 30). But the Hebrew word translated as "shone" in our modern Bibles actually has two meanings: "to shine" and "to be horned." Jerome chose the latter definition. Because his Latin Vulgate became the official Bible of the Catholic Church, Michelangelo's sculpture displays the prophet with horns.

Simple errors in translation, like St. Jerome's, can change the entire nature of a story. The childhood game of "telephone" demonstrates how quickly poor communication can undermine a true story, even as some of its major elements remain. As we consider what the geological evidence—rocks and layers of sediment—reveals about the history of our planet, there are two primary and opposing views: a slow process that took place over billions of years versus a rapid process that took place by an overwhelming cataclysmic event.

Many quickly dismiss the Bible's account of a worldwide flood and believe that the slow processes (gradualism) we observe today are what the earth has always been like. However, not only have some geologists pointed to evidence in the geologic record for a calamitous global deluge (catastrophism), but history also indicates that many ancient peoples have a shared narrative—a story about a boat, a warning, a massive flood, and a sacrifice.

Could these tales from multiple civilizations written thousands of years ago hint at a true story that a growing amount of geological research now supports? Isn't it telling that over 250 flood legends are found around the planet?

Let's look at a few of the more notable ones …

The Chinese Myth. A farmer captures a thunder god. But when the farmer goes into town, his son and daughter disregard his warning to stay away from the caged deity. Instead, they release him out of pity. In gratitude, the god warns the children there will be a great flood. He gives them a large gourd in which they are to find shelter during the deluge.

The Hawaiian Myth. Nu'u constructs a large canoe and gathers animals within it to escape a devastating Great Flood. Eventually, his vessel comes to rest atop Mauna Kea, a mountain on the Big Island. In a misunderstanding,

Nu'u attributes his deliverance to the moon and begins offering sacrifices to it. Kāne, the god responsible for the creation, descends from the heavens on a rainbow to enlighten Nu'u about his error.

The Hindu Myth. Combining the characters of Adam and Noah into one person, Manu is both the first human and the one who survives a worldwide flood. He is warned by a fish, to whom he has done a kindness, to build a boat because a flood is coming. He ties the boat to the fish's horn and is steered to a resting place on a mountaintop. After the waters recede, he performs a sacrifice by pouring oblations of butter and sour milk into the waters.

The Greek Myth. Noah's equivalent is Deucalion, the son of Prometheus, the god who created humans. When Zeus, the king of the gods, decides to destroy humanity with a flood, Deucalion is told to build an ark to save himself and his wife, Pyrrha, who is also his cousin. After nine days of flooding, the ark comes to rest on top of Mount Parnassus. The waters recede, and the couple offers a sacrifice to Zeus to learn how to repopulate the earth.

The Sumerian Myth. Perhaps the most well-known flood story outside the Genesis account is the one recorded in the poem *The Epic of Gilgamesh*. Gilgamesh is a king who, in search of immortality, meets Utnapishtim, who recounts this story of a flood: A god instructed Utnapishtim to build a giant boat to be boarded by his family, his craftsman, and all the animals. After six days and nights of flooding, the vessel lodged on a mountain. Utnapishtim then released a dove, a swallow, and a raven. When the third failed to return, he released the rest of the inhabitants. After offering a sacrifice to the gods, only he and his wife were granted immortality.

Why aren't there ancient stories of great forest fires or earthquakes ... just a worldwide flood?

Couldn't these other flood myths, however varied in detail, also hint that such a deluge really happened—drowning all people but those who boarded some kind of boat at the behest of a deity?

Considering this possibility in light of the evidence of geology in this section, there is good reason to trust the Bible record. What stands in contrast to these fantastical flood fables, with their many implausible details and cruel gods, is the Bible's portrayal of a loving deity who tries to save rebellious humanity. Even though mankind became so wicked "that every intent of the thoughts of his heart was only evil continually" (Genesis 6:5), one man "found grace in the eyes of the Lord" (v. 8).

We, too, can experience that same grace. While God promised Noah to never again destroy the world with water (Genesis 9:12–16), the Bible predicts that someday our planet will be destroyed again, this time by fire (2 Peter 3:5–7). Like He did for Noah, God promises to provide a place of safety for those who follow Him in these last days (Psalm 91:1, 2).

The Missing Sedimentary Layers

An Amazing Fact:

The largest recorded number of layers ever baked into a layer cake was 260, created in Winona, Minnesota, on September 14, 2018. The cake weighed 1,250 pounds and measured 6 feet tall, with a base of only 32 inches by 24 inches. Eleven bakers took 145 hours to create the record-breaking cake, which provided 5,000 servings.

Imagine a different kind of record-breaking multilayered material—up to 750 feet thick and spread evenly over approximately 600,000 square miles! Geologists call this giant multi-colored sediment layer of rock the **Jurassic Morrison Formation.** It blankets a large swath of the Western United States, stretching from Canada to Texas. It's also one of the best sources of dinosaur fossils.

How was the formation made? Or, for that matter, what about all the widespread sedimentary layers visible in places like the walls of the Grand Canyon? There are vast and incredibly flat examples of such deposits found worldwide.

No geological process we see today can explain the continent-wide layers being laid down so uniformly. Moving that amount of silt, gravel, sand, and rock cannot be explained by the actions of local tides, floods, or rivers. The size of particles and the huge amount of material demand something much more powerful than what we experience in the natural disasters of our day. If we accept the traditional view that these colossal layers were formed over millions of years, we immediately run into a rock wall of evidence that demands a better explanation.

For instance, one would expect a layer of sediment formed over a million years to be jagged and non-uniform, broken up by the weather or vast numbers of rivers, valleys, and canyons, along with plant and animal disruptions. Instead, the layers resemble a uniform stack of pancakes. The consistency and relative thinness of the formations challenge evolutionary explanations. The extensive distribution of these rock layers is better explained by an immense amount of floodwater not seen in our day.

One explanation for the extensive distribution of sediment layers is that a giant meteor struck the earth. However, the study of meteoritic impacts does

Definition: Sedimentary Rock

Sedimentary rock layers, also known as strata, result when sediments—particles derived from the weathering and erosion of pre-existing rocks, organic material, and minerals—are deposited in layers, typically in water environments like lakes, rivers, and oceans. Over time, these layers are buried, compacted, and cemented together to form solid rock.

not match what we see in these vast formations. An asteroid striking the ocean comes closer to explaining sedimentary layers formed by enormous amounts of water. Still, such a rapid process shoots a hole in evolution's demand for millions of years.

Missing Gaps

Perhaps the greatest challenge to the theory of naturalism, which explains how these widespread layers of sediment accumulated over millions of years, is the missing gaps between some layers based on the geologic column. Like the "missing links" in the fossil record for the evolution of different species, there are missing gaps in the sedimentary layers.

A prime example is the Hutton Unconformity located at Siccar Point, Scotland. At this sea cliff in 1788, James Hutton noticed horizontal layers of red sandstone sitting directly on top of vertical gray shales, called an unconformity. It has been assumed that, for over 20 million years, no sediments were laid down between these two layers.

Hutton Unconformity

Great Unconformity

Definition: Unconformity

In traditional evolutionary geology, unconformity is considered a long break in time in an otherwise continuous rock record. Unconformities are a type of geologic contact—a boundary between rocks—caused by erosion or a pause in sediment accumulation, followed by the deposition of new sediments.

Many geologists claim the discovery of these two layers (more accurately, the interpretation of what Hutton saw), which they believe came from different eras, proves the earth is billions of years old. Yet it is hard to believe that no sediments were deposited or that no erosion occurred in the gap between these two formations for even a thousand years, let alone 20 million. Why wouldn't more sediments continue to be laid down? And how could there not be some evidence of erosion (e.g., gullies caused by the weather) during such a long stretch of time?

Maybe there is no missing gap.

In the Grand Canyon, "missing layers" of the geologic column can be observed easily, and the missing gaps range from 6 to over 100 million years. We know these "gaps" exist because the missing layers appear in other parts of the world. Additionally, there is an unconformity at the base of the canyon called the Great Unconformity, supposedly representing a 1.2-billion-year gap!

Uniformitarianism, a basic tenet of evolution, states that rates of erosion have always been about the same. If that were true, how could the fossil record suddenly disappear in these gaps? These unconformities shout to us for a better explanation. The lack of erosion between these supposed missing gaps indicates that the layers were laid down rapidly by a catastrophic flood. If millions of years elapsed between these gaps, we would expect to find erosion from weather, evidence of plant growth, and soil formation in the lower layers next to the missing gaps.

But if we consider the biblical account of a worldwide flood, we can resolve the problem of the missing gaps. The evidence supports rapid layer formation through the movement of water. The actions of a massive flood can best explain the widespread, flat sedimentary layers.

It is worth considering what the Bible says about the conditions of the world just before the Flood. People resisted believing in a coming deluge and continued living as if nothing would change. But they were caught by surprise and "did not know until the flood came and took them all away, so also will the coming of the Son of Man be" (Matthew 24:39).

Perhaps we are about to experience another catastrophic change in our world.

Ice Core Dating: Standing on Thin Ice

An Amazing Fact:

On July 15, 1942, poor weather forced eight WW2 planes to make an emergency landing on an ice field in Greenland. All the crew were rescued, but the aircraft were eventually buried in snow and ice. Fifty years later, a P-38 Lightning was brought to the surface in pieces and restored. The plane was christened "Glacier Girl" and now flies in air shows. Incredibly, in just 50 years, some of the aircraft were covered in nearly 300 feet of snow and ice. According to traditional ice core dating methods, the planes would have otherwise appeared to be several hundred years old.

Photo by Greg Goebel

An ice core is a sample of ice typically taken from an ice sheet in places like Antarctica or Greenland. Cores are obtained by drilling with hand augers or powered drills and have reached depths of over two miles. The incremental build-up of annual snow layers means the lower layers are older than the upper ones.

Glaciologists have studied ice cores to determine Earth's ancient climate. Efforts have been made to estimate the age of the ice by counting layers and studying other elements trapped in the ice. Such predictions are known by researchers to not always be reliable. Many have dated ice cores from Greenland and Antarctica as being between 400,000 and 800,000 years old. One ice core sample found in 2016 used a potassium-argon dating method and claimed the ice was 2.7 million years old.

However, scientists raise questions about the theories used in ice-core dating. One assumption in this method is based on the **uniformitarian** view—a belief that the climate we observe today has existed over long ages with little change. The possibility that a global

flood caused catastrophic changes to the weather is discarded, and the belief in the aftermath of an Ice Age is not considered.

Another assumption is that one ice layer represents one year of accumulation. But this rejects the fact that multiple layers can accumulate within one year through storms and warming and cooling cycles. The WW2 planes found 50 years after they landed on the Greenland ice field were covered with about 300 feet of ice, indicating they should be hundreds of years old. Yet, heavy precipitation in southeast Greenland at a low altitude near the coast demonstrates how quickly ice can accumulate.

Furthermore, the deeper the ice cores, the more the weight of snow and ice above compresses and blurs the distinct layers. So scientists use other methods to estimate the age of the ice, such as the fluctuation of oxygen atoms, to determine yearly changes. But this is based on assumptions that the Earth's atmosphere has basically never changed.

If a catastrophic flood once devastated the earth, and if multiple dust-producing volcanic eruptions marked the post-flood years, the cataclysmic upheavals during this period could have nearly blocked out the sun and laid down hundreds of layers of ice in just a few decades. Changes to the earth can occur rapidly, as observed in the mudflows following the eruption of Mt. St. Helens in Washington State.

Thick ice sheets do not need millions of years to form. According to NOAA, an astonishing 95 feet of snow was recorded at Mount Baker Ski Area in Washington during the 1998/1999 snow season. That's the most snowfall recorded in a season for any location in the USA.

With enough ice and snow accumulation, thick glacial deposits can easily form in thousands of years. Counting visible bands in ice cores can work, but only up to a point. When accumulation rates are low, glaciologists must use other models to determine ages at different depths within the ice.

Perhaps it is worth considering the Bible's record of a global flood that changed the face of the earth. The deluge provides a reasonable explanation for understanding the rapid deposits of sediment layers and the accumulation of ice within a short history of the planet. In our search to understand our earth's origins, we would do well to follow the apostle Paul's advice to "test all things; hold fast what is good" (1 Thessalonians 5:21).

Mountains of Marine Fossils

An Amazing Fact:

At an altitude of 12,507 feet, Lake Titicaca is considered the highest navigable lake in the world—and the last place you would expect to find sharks. Yet students from the University of Puno in Southern Peru have discovered the remains of a prehistoric shark near the massive freshwater lake—which is not that surprising since bull sharks are still found in Lake Nicaragua.

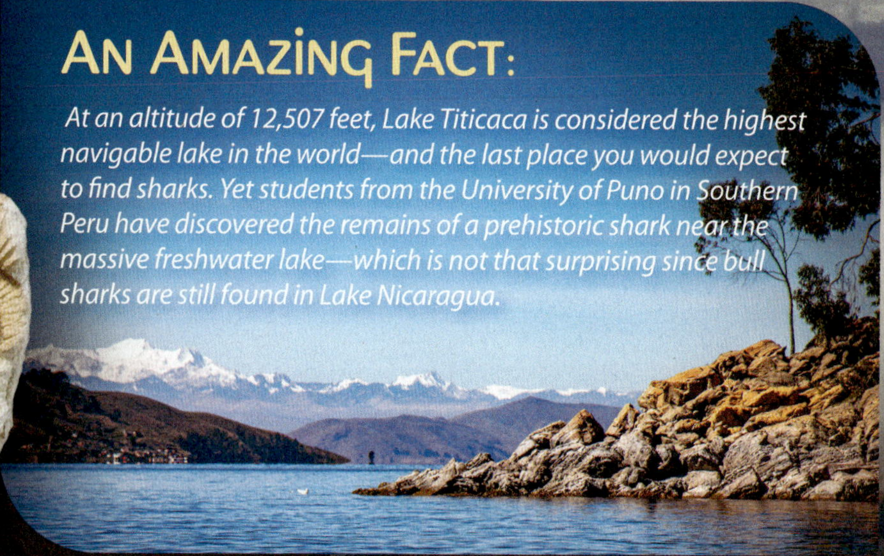

When it comes to understanding the history of our planet, fossils can give us a few insights. Fossils are the preserved remains (or impressions) of organisms said to have lived more than 10,000 years ago. Paleontology is the study of fossils, their age, how they formed, and, for many scientists, their evolutionary significance. Using radiometric dating, a technique with serious limitations (see p. 62), the oldest fossils are dated to about 3.5 billion years.

Curiously, not all "ancient" marine fossils are of extinct sea creatures. Some species still roam our oceans today—including the chambered nautilus, Atlantic horseshoe crab, sea lily (crinoid), coelacanth,

▲ *Samples of ammonite fossils found in Madagascar.*

Even in the world's highest mountains, the Himalayas of Nepal, ammonite fossils (squids with coiled shells) have been discovered in their limestone layers.

and the goblin shark. These are called "living fossils." But is it true that they have lived in our oceans for hundreds of millions of years in a state of suspended evolution?

Many paleontologists believe the fossil record supports a slow evolutionary process of life developing over millions of years. However, those who believe in the biblical record of a worldwide deluge see evidence of catastrophic death. One of the curiosities bolstering the theory of a global flood is the discovery of massive amounts of marine fossils on mountains.

Did you know that 95 percent of all fossils are actually marine fossils? And sea animals without a backbone, known as marine invertebrates, compose the largest number of fossils we encounter. Their location in sedimentary rock provides evidence that a catastrophic flood of water buried them rapidly.

Marine fossils are rarely located at the bottom of the ocean. Instead, they are often found on continents and high mountains. How did they get there? Ocean waters must have risen to flood the land and mountains. That's exactly how the Bible describes the deluge. "And the waters prevailed exceedingly on the earth, and all the high hills under the whole heaven were covered" (Genesis 7:19).

Geologists agree that fossils of sea creatures are found on every continent in layers of rock high above sea level. Most of the rock layers in the Grand Canyon contain marine fossils, including the topmost layer at the canyon rim, which is between 7,000 and 8,000 feet above sea level.

Evidence of tectonic plate shifts in the continents, which caused the land to buckle, reveals that mountains before the deluge didn't reach today's elevations. So the flood waters didn't need to rise to the height of Mt. Everest at 29,032 feet. This means the layers of sediment in which we find marine fossils were laid down at the time of the Flood, after which the layers buckled and lifted. See Psalm 104:8.

The massive number of marine fossils buried haphazardly on the continents indicates a global flood. They are silent witnesses to a catastrophic event that carried billions of sea creatures onto the land and buried them all over the earth. Such evidence supports the Bible record!

Finding marine fossils on mountainsides is not a recent phenomenon. The Greek philosopher Xenophanes (c. 570–478 BC) found seashells on mountains and guessed that the world experienced a colossal flood that buried the living creatures in sediment.

The Geologic Column: A Modern-day Conundrum

An Amazing Fact:

The Grand Canyon is one of the most studied landscapes in the world. It is a mile deep, 277 miles long, and 18 miles wide—measuring 1,904 square miles. In comparison, Rhode Island is 1,214 square miles. Nearly 6 million people visit the Grand Canyon each year and are told this natural wonder took 6 million years to carve out.

What could be more helpful than to walk backward through time and observe the geologic history of Earth? The Grand Canyon's Trail of Time path invites you to "travel back millions of years" via a 2.8-mile walk along the South Rim. Every large step you take represents a million years, and markers on the trail, along with interpretive signs and displays, depict the theory of a long, slow process to eventually create this stunning gorge.

That's the claim.

But does this explanation match the actual evidence found in the rocks? Do all the layers of sediment and fossils found in the rocks, which are said to create a "geologic column" of Earth's ancient beginnings, stack up? Do these layers point to a slow evolutionary process for life arising on our planet—or to a relatively recent global catastrophe that utterly changed the face of the earth?

The Geologic Column

The geologic column is a visual representation created by geologists of the rock layers that make up the earth's crust—of which the Grand Canyon in Arizona is held up as a prime example, even though some layers are missing. Scientists have collected data from many different places around the globe and blended them together into a composite picture of our planet. According to evolutionists, these multiple layers of rock and sediment reveal the 4.5-billion-year geological history of Earth.

This column is important in supporting the theory of evolution, which requires millions of years for life to have evolved. If one accepts the assumption of a slow evolutionary process, the fossil record must fit the presupposition of long ages.

However, geologists admit they use circular reasoning when assigning dates to the fossil record. That is, the geologic column is used to support the age of fossils, but fossils are also used to support the age of the rocks in the column.

Cracks in the Column

Multiple factors raise questions about the evolutionary understanding of the geologic column. First, it's helpful to know that no place on Earth features the complete column; it appears only in textbooks.

Second, rocks do not come with dates stamped on them, so the different layers in charts of the

column have been assigned time frames based on the presuppositions of evolutionary theory, which claims that life evolved from simple forms to more complex ones. Yet the first fossils appearing in the Precambrian show high levels of complexity and diversity, which does not align with Darwin's theory.

If life forms evolved incrementally over millions of years, the fossil record should be full of transitional fossils between the life forms found in the geologic record. However, such fossils are not found—even 150 years after Darwin predicted their existence. (See also pages 10 and 29.)

Evolution's Geologic Time Scale

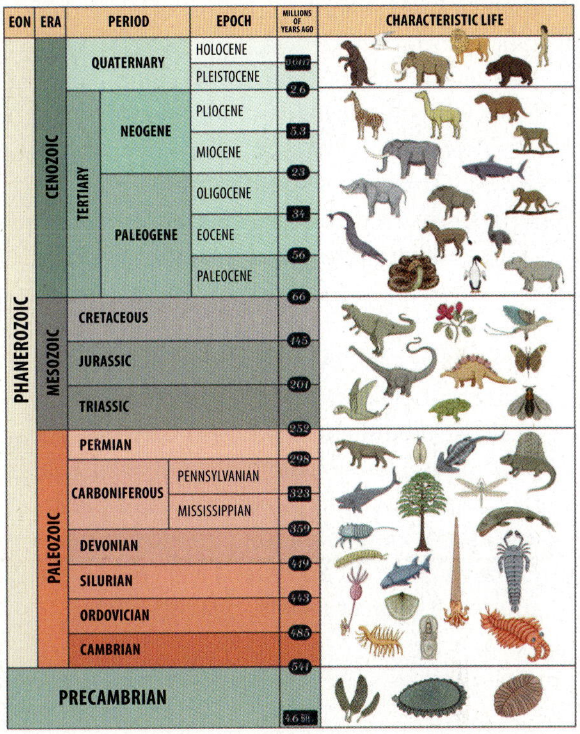

Example of Geologic Time Scale. This illustration is a composite made from many different locations around the world.

Uniformitarianism vs. Catastrophism

Uniformitarianism is the theory that the earth has always changed in uniform ways and that the present is the key to the past. It is the assumption that governs all evolutionary theory. **Catastrophism** states that changes in the geologic record are the result of events that took place at rates that were dramatically higher than what we observe today—which would include the Genesis worldwide flood.

Perhaps some of the most visible challenges to the uniformitarian theory, which teaches that layers of sediment and rock were laid down over hundreds of millions of years, are the many examples of bent (or folded) layers of rock and sediment. If these layers were indeed laid down over a long period of time, we would expect them to harden into rock. Yet, many examples show strata (some 4,000 feet thick) that had to be soft to bend at radical angles.

However, the evidence of these folded rocks matches that of a catastrophic

Earth's crust, the outermost layer of the planet's solid surface, is composed of various rocks, minerals, and geological formations. The crust is thin compared to Earth's other layers, with an average thickness of about 22 miles on land. Curiously, 75 percent of the crust is composed of sedimentary rock, the formation of which requires water.

flood, such as the one we read about in Noah's day.

Another feature that invalidates the geologic column being laid down over hundreds of millions of years is ripple marks in sandstone, which shows these layers were covered up quickly. If they were laid down over millions of years with each layer exposed to millennia of weather erosion, these marks would have been eroded. There is also a noticeable absence of structures, such as animal burrows and plant roots, crossing through layers of sediment, again indicating rapid burial. Even the significant lack of erosional features from weather indicates the layers were laid down rapidly, as would be found in the global flood described in the Bible.

The creation of the Grand Canyon, often held up as a prime example of uniformitarianism, has been explained as being made by a small amount of water over long periods of time. However, the vast and even layers of sediment suggest it was actually made with a lot of water over a short time frame. A modern-day example of another canyon that formed quickly is **The Little Grand Canyon** at Mount St. Helens.

Folded limestone
Crete, Greece

Photo by Rygel, M.C.

Folded calcite and gypsum
Texas/New Mexico state border

Ripple marks in sandstone
Watarrka National Park, Australia

The Little Grand Canyon

On March 19, 1982, an eruption at Mount St. Helens drove a 20-mile-long mudflow westward, carving out a brand-new canyon measuring140 feet deep—roughly 1/40th of the Grand Canyon, hence the moniker, "The Little Grand Canyon." This dramatic event witnessed by modern scientists challenges the evolutionary assumption that the Grand Canyon had to have been formed over millions of years by a small amount of water.

In the New Testament, the apostle Peter warns of the destruction of the world in the future by pointing to the flood that destroyed the world in Noah's time. He describes scoffers who say, "Since the fathers fell asleep, all things continue as they were from the beginning of creation" (2 Peter 3:4). But then he reminds his readers that all things have *not* continued since the creation. There was a disruption since the beginning when "the world that then existed perished, being flooded with water" (v. 6).

There is abundant geologic evidence for a worldwide flood that destroyed Earth in the past. The rapid burial of plants and animals, the vast stretches of uniform sediment found around the world, and the folded layers that show rapid changes by water are just a few reasons to believe the Bible record is true.

The Cosmos

Ex Nihilo: Out of Nothing

An Amazing Fact:

Our solar system exists on an arm of a spiral galaxy known as the Milky Way, which contains more than 100 billion stars and is spinning at roughly 168 miles per second—that's about 600,000 miles per hour. Our solar system is also approximately 26,000 light-years away from the galactic center, meaning it will take at least 225 million years for our solar system to rotate one time around the center of the Milky Way!

From where did our massive universe come—not just the mass and energy, but the immensity of space itself? How do we even begin to explain its existence?

There really are only two competing explanations:

1. The universe came into existence by *natural* processes.
2. The universe came into existence by *supernatural* processes.

A *natural* process is one that is attributed to purely naturalistic means without the intervention of a sentient being. A *supernatural* process is one that occurs outside of nature and beyond its regular laws.

These two alternatives cover the entirety of possibilities for the beginning of our universe. For each one, there are many possible variations.

For instance, most religions teach that the universe began with the actions of a supernatural being or beings. There are also many theories to explain the existence of the universe from natural processes, including the **Big Bang theory** and the **steady-state theory**.

Many scientists today publicly reject supernatural explanations outright. They assert that such processes cannot be investigated scientifically, so they reject any theory that requires a supernatural component that cannot be tested. Instead, they focus their energies on discovering natural processes that can explain the creation of our universe.

But what if the natural explanations are inadequate?

After all, scientists have been exploring this particular field for centuries. And while they have learned many valuable things about how our universe functions and the natural laws that govern

The Steady-State Theory

In the past, many scientists believed that the universe has always existed, maintaining its density as new matter continually forms at the same rate that old matter disappears. They believed that proving such a claim would show no need for a beginning act of creation or a supernatural creator. While eventually dismissed in favor of the Big Bang model, the steady-state theory is resurging in popularity among some scientists who see mounting problems with the Big Bang.

No Time for Infinity?

The Kalam cosmological argument posits that the universe cannot have existed eternally in the past because if it did, we would never reach the present moment. The reasoning is that traversing an infinite amount of past time would make it impossible to arrive at "now." Since we have indeed reached the present, this suggests that the universe had a definite beginning.

it, they still have no idea how natural processes could have produced the universe, its raw materials, and the complex life forms we see on Earth. Furthermore, they cannot explain where the natural processes came from in the first place!

Ultimately, they cannot explain how *something* can come from *nothing*.

Instead, scientists today are appealing to new theories that cannot be tested or measured: multi-universe theory, string theory, and dark energy and matter. None of these can be observed or measured with instruments we now possess, but many scientists embrace them as possibilities anyway—demonstrating a form of faith. How can these theories be scientific if the scientific method cannot be used to support or falsify them? For that matter, how can they be called "natural"?

Faith and Science Are Not at War

Many scientists reject the theory of a supernatural origin because they claim it falls outside the domain of science. But does it? Here are a few reasons why science affirms a supernatural creation of our universe.

#1 The Universe Had a Beginning

One of humanity's initial beliefs was that the universe has always existed. Secular scientists often embraced this idea because it dismissed the need to explain origins at all. However, as their tools advanced, their ability to perceive and measure the universe

was enhanced, and we soon discovered a startling fact: *The universe is expanding!*

Point A to Point B: Traveling Galaxies

As mentioned in the introduction, in the 1920s, scientist Edwin Hubble was surveying the night sky with what was then the largest telescope in the world. He discovered that some of the dots of light were not actually stars but instead entire galaxies—each with hundreds of millions of stars!

While he continued to observe, he noted that these galaxies were moving away from each other at astonishingly high speeds. Using this data, he reasoned that if we could rewind time and reverse the trajectory, we would see the galaxies and all the matter in the universe converging on a single point in the distant past—meaning that the universe could not have existed before that point.

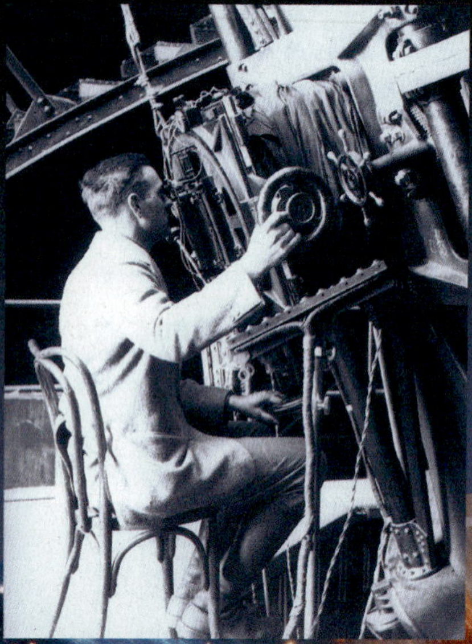

Edwin Hubble

This finding would ultimately lead to the theory known as the "Big Bang," one of the most essential components of cosmic evolutionary thought, one that concedes that all matter and energy in our universe had a starting point. While the Bible does not call it the Big Bang, it does say our cosmos had a "beginning" (Genesis 1:1).

Dying Stars

Did you know that scientists once believed that the sun was immortal? However, as our ability to measure and understand stars increased, it became evident that these giant balls of burning fuel follow the same laws of nature as all other matter in our universe. The lifecycle of the stars,

The Limits of Physics

While the natural laws of physics explain how the observable universe operates, they cannot be used to formulate an explanation for the creation of the universe itself.

expending energy and ultimately dying, is another evidence that they, just like our universe, had to have a beginning.

#2 A Thing Which Does Not Exist Cannot Create Itself

Popular scientist Stephen Hawking wrote in his book *The Grand Design*, "Because there is a law such as gravity, the Universe can and will create itself from nothing. Spontaneous creation is the reason there is something rather than nothing, why the Universe exists, why we exist."

Yet spontaneous creation is at odds with basic logic:

> There is no case known (neither is it, indeed, possible) in which a thing is found to be the efficient cause of itself; for so it would be prior to itself, which is impossible (*Summa Theologiae* I.2.3).
>
> If, then, something were its own cause of being, it would be understood to be before it had being—which is impossible (*Summa Contra Gentiles* I.22.6).

If the universe could not have created itself, that means something *outside* the universe must have been the cause of it. And if something outside the universe created the universe, it would be beyond our ability to measure or detect it with our senses or scientific devices.

The fact is, no matter what you believe about the beginning of the universe, it is going to involve elements that cannot be measured or defined by science. Either a process beyond the realm of our understanding caused the beginning of all energy and matter by accident—or a supernatural creator did it on purpose. Since we cannot witness the beginning of this universe, we can only infer its origin from what we observe now.

The Supernatural Origin of Our Universe

It is one thing to believe that the creation of our universe involved forces that science cannot explain—it is quite another thing to understand what those forces are. The only way we could ever know the truth about the beginning of our universe is to be given testimony from a source that was there when it happened—a being who exists outside of our universe. Without such testimony, the universe's origin is mere philosophy.

We are curious creatures, a species that can reason logically. To believe in some unknown "supernatural" process or entity is to reach an unresolved conclusion. The good news is, if we honestly look further into our origins and explore beyond our philosophies, we find evidence that our universe must have been created by a divine being—One whom each of us can know personally.

SOURCES
https://creation.com/universe-had-a-beginning
https://gizmodo.com/mathematic-proof-that-the-universe-had-a-beginning-5904714
http://arxiv.org/abs/1204.4658

Serpens Nebula from NASA's James Webb Space Telescope

The Immensity of the Universe: The Power of God

An Amazing Fact:

If you were to travel from Earth to the sun at the speed of a typical passenger jet (approximately 550 mph), it would take about 19 years of continuous travel. A trip to Pluto would take about 760 years. If the sun were the size of an elephant, the relative size of the Earth would be like a plum, orbiting about a quarter mile away, and Pluto would be the size of a pea, located about eight miles out. Yet, when compared with the scale of the universe, our solar system is tiny and impossible for the human mind to truly grasp.

It's futile to explain the scope of the cosmos using everyday systems of measurement. And with each advancement in astronomy, new powerful telescopes are peering deeper than ever into the heavens, leaving us even more humbled as we consider our minuscule spot in the vastness of space.

For example, the sun may seem not that far from Earth, but it is 93 *million* miles away. And Alpha Centauri, the next closest star system, is 25 *trillion* miles away.

On a clear night, with no light pollution to block your view, it appears to the naked eye that you can see tens of thousands of stars. But you can actually see only about 2,000. A good pair of binoculars increases that number to perhaps as much as 100,000. But you still haven't seen anything yet!

Our Milky Way galaxy alone has between 100 and 400 billion stars. It's about 100,000 light years in diameter, and because of our location within it, we can see only a small percentage of the stars.

When the Hubble Space Telescope was launched in 1990, some astronomers doubted it would provide radically new facts about the universe. But in 1995, after spending 10 days focusing the Hubble on a seemingly empty pinpoint in the sky, they were stunned by what they saw—a picture of 3,000 never-before-seen *galaxies*. Known as the Hubble Deep Field, the image revolutionized the way

▲ **HDF Details**

◀ **Hubble Deep Field**

Hubble Ultra Deep Field

HUDF Details

we understand the universe. Yet in 2004, the Hubble Ultra Deep Field image revealed nearly 10,000 galaxies.

The revelation was astonishing. Further deep-field observations unveiled a new picture of the immense size of our universe. It is filled with hundreds of billions of galaxies, each containing multiplied billions of stars. Imagine holding a grain of rice at arm's length up against the night sky and then realizing that hiding behind it are billions of stars! Thanks to the Hubble, the "observable" universe expanded to include an estimated 100 billion galaxies.

Our view of the universe became even more mind-blowing after the new James Webb Space Telescope was launched on December 25, 2021. A hundred times more powerful than the Hubble, its mission was to look even deeper into the universe. In July 2022, the first set of its deep-field images was released, showing the beauty of the heavens in stunning resolution.

Too Many Zeros!

As a shorthand to explain the vast expanse of space, astronomers use "light years" to communicate distance. The distance light can travel in one year is 5.88 trillion miles—meaning Alpha Centauri is 4.37 light years away.

The new images have reinforced the idea that the stars in the universe may outnumber the grains of sand on Earth.

Interestingly, God told Abraham that the stars of heaven are as numerous as the sand on the seashore (Genesis 22:17).

The Scriptures also tell us that observing the universe will teach us about the Creator:

> The heavens declare the glory of God; and the firmament shows His handiwork. Day unto day utters speech, and night unto night reveals knowledge. There is no speech nor language where their voice is not heard. Their line has gone out through all the earth, and their words to the end of the world (Psalm 19:1–4).

Our knowledge of the universe is incomplete; there is much we do not know. But the stunning vastness of what we do see should cause us to speak with humility about the little we know and encourage us to listen to the message about an infinitely powerful Creator.

The Big Bang's Big Trouble

An Amazing Fact:

The James Webb Space Telescope is one of the world's most technologically advanced marvels. It took 26 years for this highly sophisticated instrument to move from concept to fully operational. The telescope can magnify and capture images of objects 100 times dimmer than those captured by the Hubble Space Telescope. The James Webb orbits a million miles from Earth and must be kept below –370 degrees Fahrenheit to function properly. What do its new stunning images of the heavens tell us about the beginning of the universe?

When Christians are asked about the origins of the universe, their answer can sound quite simple: "In the beginning God created the heavens and the earth" (Genesis 1:1). Of course, the next question is usually, "Where did God come from?" However, the Bible explains that God had no beginning.

"From everlasting to everlasting, You are God" (Psalm 90:2). God is the only eternally independent, self-existent One. This is undoubtedly a mystery—but it's not any more challenging than the mystery faced by atheists: Where did the universe come from? How did something come from nothing?

For nearly a century, their answer has been "The Big Bang."

What Is the Big Bang Theory?

In 1929, astronomer Edwin Hubble helped prove the universe was expanding by identifying a "red shift" of light in stars and galaxies. A red shift occurs when light waves are stretched toward the red side of the spectrum, which happens when a light source is moving away from the observer.

In theory, you could reverse calculate to estimate an approximate beginning point and time for the universe. Astronomers thus began to theorize there had been some kind of "explosion" billions of years ago from a point in space. The famous astronomer Fred Hoyle was skeptical of such an event and jokingly referred to the idea as a "Big Bang." However, the name stuck, and even Hoyle eventually believed it.

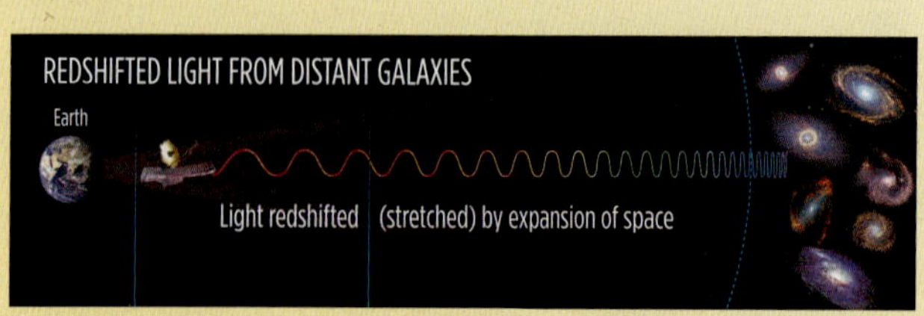

Furthermore, in 1964, two radio astronomers, Robert Wilson and Arno Penzias, were using a large horn antenna to listen to the sounds of the universe. They reported a humming coming from all directions, which seemed to be additional supporting evidence for the Big Bang. It was interpreted to be the energy from that explosion.

Researchers believed this theory was supported when the WMAP satellite was launched in 2001; it was designed to map out and study this sound. What it found was the so-called "Cosmic Microwave Background." Scientists thought they could now actually see the energy of the Big Bang's aftermath.

A Theory in Crisis?

What happens when a new discovery conflicts with a long-held theory? Good science would suggest that you move on to a better model. Well, since the latest data has been coming in from the James Webb telescope, the Big Bang Theory, one of science's most popular theories regarding the universe's origin, has been turned upside down.

The Hubble Space Telescope, launched in 1990, allowed astronomers to see objects nearly 13 billion light years into space. A light year is the distance light travels in one Earth year—about 5.88 trillion

Webb's First Deep Field

miles! But because the farthest images from the Hubble were very faint, a more advanced telescope was needed to see further into the universe's past—a universe estimated to be about 13.7 billion years old.

Thus, after nearly 30 years of development, the James Webb Space Telescope was launched in 2021 with the hopes of seeing considerably further "back in time. "What did it reveal? In detail much clearer than the Hubble's, it brought to light stunning images of galaxies that should have looked young, being near that 13.7 billion light-year threshold. But instead of young galaxies,

Wilson and Penzias with the Holmdel Horn Antenna

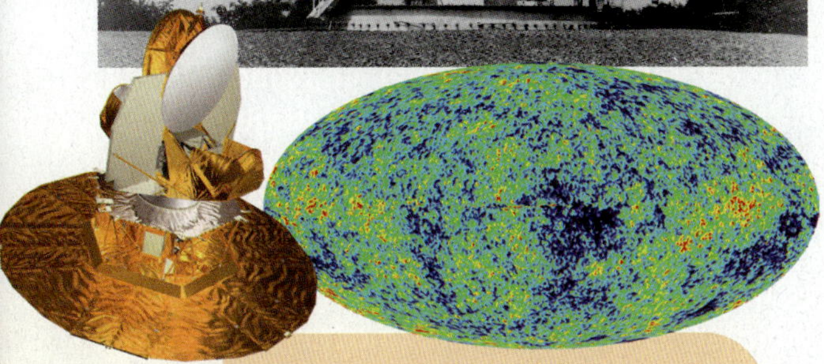

Cosmic microwave background imaging from the WMAP satellite

astronomers found mature ones, which would've been impossible had there been a Big Bang that long ago. Interestingly, just like in the geological fossil record that lacks transitional fossils, they didn't find transitional forms of galaxies but rather fully formed ones in every direction. Many world-famous astronomers are now saying, "It's back to the drawing board!"

"Do we really understand the early phases of the formation of these galaxies?" said Garth Illingworth, an astronomer at the University of California, Santa Cruz, at a press conference held by NASA. "This has posed a lot of questions for the theorists."

Many physicists have always known the Big Bang theory was in trouble from the beginning because it violates the First Law of Thermodynamics: "Energy cannot be created or destroyed." Put simply, you cannot get something from nothing. For there to be a Big Bang, there had to be something, some event or cause, that would trigger this cosmic explosion.

Creationism and the Red Shift

Thanks to the red shift, the universe *does* seem to be expanding.

For years, scientists didn't like the idea of a great creative act like an explosion, preferring to believe that the universe had always existed in a sort of "steady state." With new data coming in from the James Webb, it will be interesting to see what new theories will arise to explain the universe's origins.

Christians have often said, tongue-in-cheek, "I believe in the Big Bang. God spoke, and BANG, it happened!" However, Harvard biologist Richard Lewontin once spoke for many when he said, "We must not allow a divine foot in the door!" In so stating, he revealed the powerful atheistic foundations for much of modern science.

Evidence points to an infinite God who supernaturally created an infinite universe with countless galaxies. As more powerful telescopes look deeper into space, it brings to mind the words of the psalmist:

> When I consider Your heavens, the work of Your fingers, the moon and the stars, which You have ordained, what is man that You are mindful of him, and the son of man that You visit him? (Psalm 8:3, 4).

SOURCES

O'Callaghan, Jonathan. *Scientific American,* December 1, 2022.

New York Review of Books, January 9, 1996.

The Moon:

Earth's Life-Sustaining Partner

An Amazing Fact:

The sun is 400 times larger than the moon. It also happens to be 400 times farther away. That means the sun and moon appear around the same size in the sky. This unusual occurrence allows for spectacular solar eclipses, which permit scientists to study the sun's corona, the outer ring of fire surrounding the sun when the disk of the moon darkens it. Was this perfectly sized natural satellite formed by accident or created by design?

We often get romantic when enjoying the stunning beauty of a full moon. Its gentle blue light can quiet the soul. Early skygazers first called Earth's partner "the moon," thinking it was the only moon in our solar system. It wasn't until AD 1610 that Galileo discovered four moons orbiting Jupiter.

But there's more to this orbiting friend than serving as a quiet night light to our planet.

Of the 290 moons in our solar system, our moon is the fifth largest, but its relative size to the earth is quite large. Indeed, no other planet in our solar system has a moon like the earth.

The moon plays several vital roles in sustaining life on our planet. It stabilizes the tilt of our orbit, moderating the degree of wobble and causing seasonal changes. Earth would rotate much faster without the moon, causing winds over 100 miles per hour and creating 8-hour days. Without our massive moon, the gravitational forces of the sun and other planets would cause the earth's tilt to change dramatically over the years, which would have a devastating impact on the climate and life.

The orbit, size, and distance of our moon are perfect for creating ocean tides that prevent the earth's water from stagnating. Daily tidal changes give life to many sea creatures and cleanse the beaches. Biologists have also discovered how the moon aids animals in mating, hunting, and communicating. Birds use the moon as a map to guide their flight plans during migration.

The moon's magnetic field also shields the earth from the effects of the sun's radiation and helps to protect it from damage caused by asteroids, meteorites, and comets. Just look at all the craters that pockmark the lunar surface!

The formation of lunar craters has been debated since Galileo first noticed them in 1609. Some thought they were created by volcanic eruptions, but astronomers agree that the craters are impact sites made in seconds and did not form over eons. On March 17, 2013, an 88-pound meteoroid struck the moon's surface at 56,000 mph—the impact was visible to the naked eye.

March 17, 2013 impact crater

Our moon is indeed unique and special. Is it because it was made for a purpose? Notice what the Bible says about the fourth day of Creation:

> Then God said, "Let there be lights in the firmament of the heavens to divide the day from the night; and let them be for signs and seasons, and for days and years; and let them be for lights in the firmament of the heavens to give light on the earth"; and it was so. Then God made two great lights: the greater light to rule the day [the sun], and the lesser light to rule the night [the moon]. He made the stars also. God set them in the firmament of the heavens to give light on the earth, and to rule over the day and over the night, and to divide the light from the darkness. And God saw that it was good. So the evening and the morning were the fourth day (Genesis 1:14–19).

It seems like much more than a coincidence.

More Interesting Facts About the Moon

The moon is an average of 238,855 miles away from us, so about 30 Earths could fit between us and the moon. It is in synchronous rotation with Earth, meaning the same side always faces Earth. The moon has no atmosphere to protect it from cosmic rays and meteorites—which leads to huge temperature fluctuations, averaging from 250 to minus 250 degrees Fahrenheit.

SO MANY CHOICES!

afbookstore.com

Find hundreds of resources on biblical and lifestyle topics!

◀ Scan the code to visit our online bookstore!

Myths

Human Missing Links Still Missing

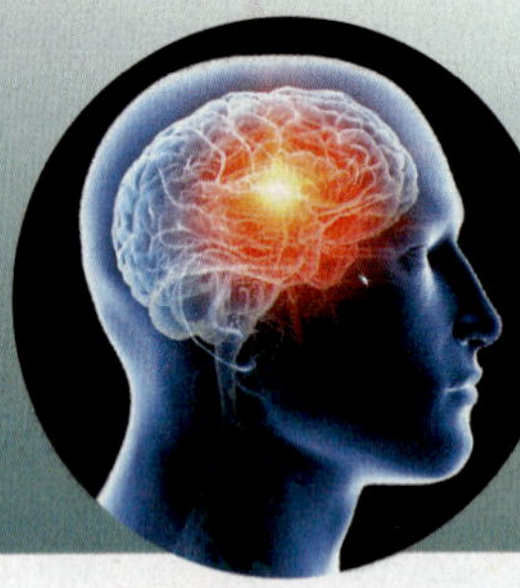

An Amazing Fact:

A hemispherectomy is a rare surgery that either removes or disconnects half of the brain from the other half. It is generally reserved for severe cases of epilepsy. Incredibly, children who experience this procedure can still live a normal life with minimal handicaps. This is because the human brain has the incredible ability to reorganize and reassign functions.

No other visual has done more to shape the public's understanding of human evolution than "The March of Progress," created by artist Rudolph Zallinger for Time-Life Books in 1965. Originally titled "The Road to Homo Sapiens," the foldout picture shows a sequence of 15 primate figures walking as if in a parade line.

The far-fetched illustration suggests a somewhat smooth process of 25 million years of evolution from a gibbon-type, primate-like common ancestor to modern man.

But there are massive differences between apes and humans today. Just the complexity of the human brain—its size and functions—inexorably separates humans from gorillas. Though evolutionists suggest that humans are the result of a purposeless natural process, they cannot explain how the human mind could have randomly evolved.

Despite the herculean effort made by paleontologists, the woefully inadequate fossil record of ancient prehumans casts a shadow on "The March of Progress." Skeletons of "early man" are

The original March of Progress illustration from *Early Man* (1965). Fair use.

Impossible Survival

Humans are one of the few species that remain helpless for years after their birth. A baby gazelle can stand and walk within hours of birth and run in a couple of days, but a human can't walk for 12 months. Many animals are born naked but quickly grow a coat of fur. Humans remain naked their entire lives and depend on the fur of other animals. How did humans, which would seem to be the most unlikely to survive, take over the world?

often little more than plastic models based on biased imaginations trying to fill the gaps.

However, why have literally thousands of dinosaur fossils, hypothetically 65 million years old, and some virtually complete skeletons been recovered? By contrast, only a meager handful of bone fragments of supposed primitive man have been brought forward. But are these ape-like species forerunners of modern humans—or simply an unknown extinct species of ape?

If humans have existed for 500,000 years, why is it that evidence of human activity goes back only several thousand years? The ancient Sumerians of Mesopotamia are considered the oldest civilization in the world yet have been dated to only 4,000 BC. How could the slow evolution of humans leave behind so little evidence of past activities?

And how could any species virtually leap from the trees or crawl from caves dragging their knuckles, and then suddenly, in just a few thousand years, begin to understand geometry, engineering, and astronomy and start building pyramids? The complexities of the human mind in performing complex mathematical equations, creating rockets that carry people to the moon, and composing beautiful symphonies remain an enormous challenge to naturalistic theories of human origins.

Why did the inexplicable driving force that led humans to expand their intelligence leave all other creatures behind? How do people alone have our unique power of speech, reason, and choice? Why are we the only creatures capable of moral responsibility and limitless imagination? All this evidence points to an intelligent design; humans are formed after a divine pattern, not a purposeless evolution.

"What a piece of work is a man! How noble in reason, how infinite in faculty! In form and moving, how express and admirable! In action how like an angel! In apprehension how like a god! The beauty of the world! The paragon of animals!"—William Shakespeare

The Fossil Record: Gaps and Gaffes

An Amazing Fact:

Some species living today have allegedly existed for millions of years yet are essentially the same as their supposed ancient ancestors. These are called "living fossils." For instance, evolutionists say the first dragonflies appeared 300 million years ago. Yet somehow, they stopped evolving and have remained virtually unchanged from fossil specimens all this time.

Charles Darwin built his theory on the assumption that Earth's many different life forms had evolved over eons, gradually transitioning from one species to another better suited for the environment. He proposed that all these living organisms came from a single common ancestor. If true, those gradual changes to plants and animals should be clearly reflected in an innumerable amount of fossils.

But even in his day, Darwin was forced to admit that the **fossil record** didn't support his theory. However, he assumed that those elusive transitional fossil forms would be discovered over the next few decades and that the fossil record would vindicate his theory.

While surveying flora and fauna on the Galapagos Islands, Darwin observed small variations in finches, such as in their beak sizes. But the finches, though varied in color, size, and shape, were always just finches. Nor is there any fossil evidence that finches were ever anything but finches. If macroevolution has occurred in the past, the evidence ought to be abundant in the rocks—but it is conspicuously missing.

While a few examples of transitional fossils have been brought forward, they have only revealed that the finders were mistaken. (See sidebar.)

Rather, the fossil record supports only the sudden appearance of fully complete organisms and the sudden disappearance of others. There are no examples of partial, in-progress developments—only partial fragments of highly complex, fully mature creatures. The Cambrian explosion (p. 10) is a massive nail in the coffin of the theory of evolution, as nearly all major animal phyla appear at once, supporting the Bible's claim that everything was created around the same time.

The trilobite is an excellent example of an extinct marine arthropod that suddenly shows up as a complex creature without transitional fossils. Many flying and crawling insects, like dragonflies, also suddenly appear, fully formed with extremely sophisticated aerodynamic abilities, yet no ancestors are in the fossil record. Not only do organisms

Fossilized trilobites

Mistaken Transitional Fossils

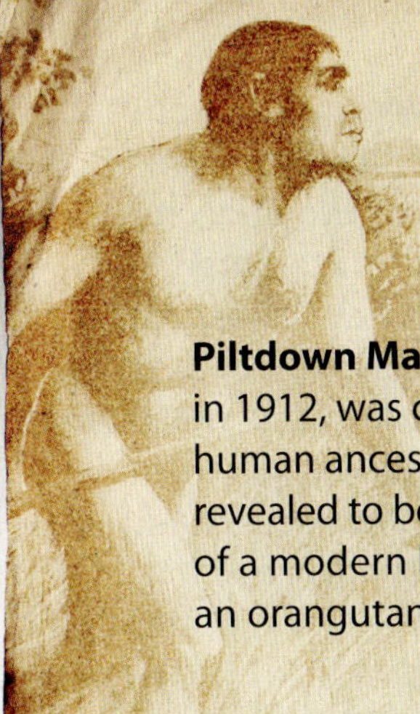

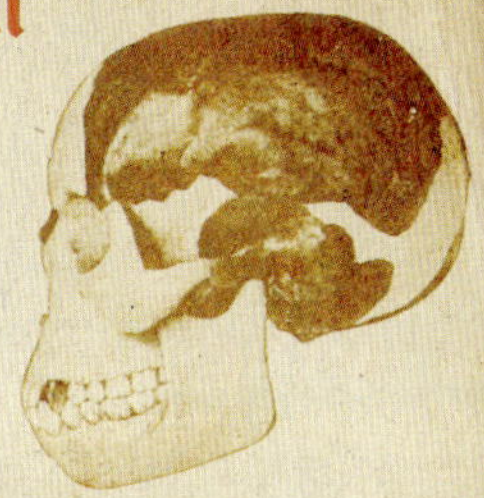

Piltdown Man, "discovered" in 1912, was considered a key human ancestor. In 1953, it was revealed to be a hoax, consisting of a modern human skull and an orangutan jaw.

Nebraska Man, a fossil based on a single tooth found in 1922, was initially thought to belong to a human ancestor. Later analysis revealed it belonged to an extinct species of peccary, an animal related to pigs.

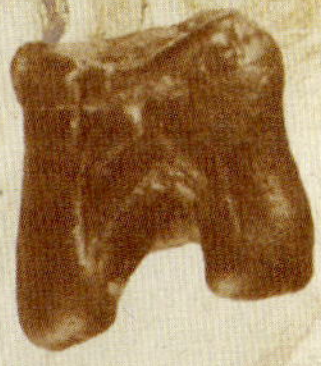

Ida was discovered in 1983 and heavily publicized in 2009 as a "missing link" in primate evolution. Subsequently, it was proven not to be a direct ancestor of humans but rather a representative of an early branch of primates.

Ida photo by Jens L. Franzen

suddenly appear, but they remain unchanged for long periods of "geologic time" before going extinct, leaving no examples of intermediate successors.

Microevolutionary changes undoubtedly occur within these groups, like desert hares turning into Arctic hares, but there is no evidence that worms evolved into wart hogs or crabs into cats. Though adaption within types can happen, no evidence exists that evolutionary intermediates existed between major groups.

The sequence we see in the fossil record shows us how organisms were buried at the time of the Flood, not how they originated. It does not show transitional "missing" links between the different types throughout the **geological column.** The rocks do not support gradual evolution but point to God's creation and, later, a global flood.

Darwin's theory breaks down under these boulders of evidence. But according to the Bible, all major types of life were created as distinct kinds in six days: "And God made the beast of the earth according to its kind, cattle according to its kind, and everything that creeps on the earth according to its kind. And God saw that it was good" (Genesis 1:25).

Homology: Common Assent ... or Common Designer?

An Amazing Fact:

Bottlenose dolphins are legendary for their ability to travel through the seas silently, at great speed, and with minimal effort. Because of this, the U.S. Navy has invested significant time and money in analyzing the swimming technique of dolphins to improve submarine design and performance.

What do cats, camels, and crocodiles have in common? They each have two eyes, two ears, four legs, and a heart—and they breathe air. One of the foundational arguments for evolutionary theory is this **homology,** which means that certain animal similarities prove they evolved from a common ancestor.

Charles Darwin wasn't the first to notice such similarities. In fact, scientists before Darwin's time argued that these similarities indicated a common *designer*, not a common ancestor.

Many books supporting evolution today often display a picture of the similarity of the arrangement of parts from different animals, such as dogs, bats, whales, and humans. The forelimbs of these vertebrates are color-coded to show a basic pattern. This proves, it is argued, that they evolved from each other or a common ancestor. Evolutionists have also attempted to demonstrate other types of similarities, such as the cells in living organisms, hereditary information based on the same genetic code, and the similarity of comparable DNA sequences.

However, scientific research indicates that homology rests on shaky ground.

What first appears to be a homologous structure is shown to be quite dissimilar upon closer examination. Organs in one species are sometimes developed by a completely different embryogenic

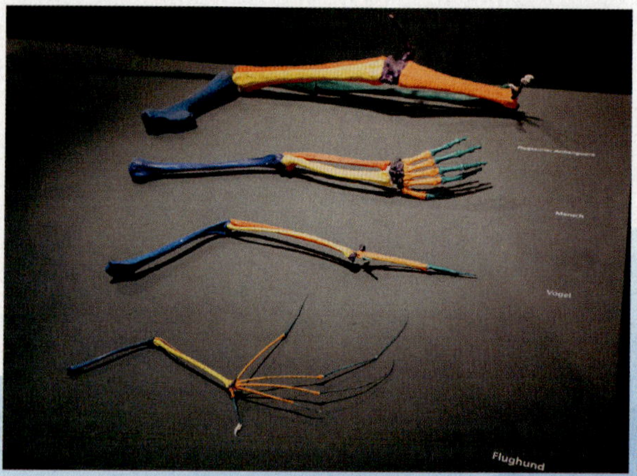

EdNurg - stock.adobe.com

route than in another. Different genes direct growth in other species in various ways. Moreover, these same similarities can be used to argue in favor of a Creator God who designed a useful pattern that works. If the requirements for living on this planet are the same for many creatures on land, in the air, or underwater, wouldn't it make sense that there would be a similarity between organs, lungs, or bones? Why would a divine engineer need to start from scratch when making different types of birds, fish, or land animals?

Consider the variety of vehicles traveling on highways worldwide—cars, motorcycles, pickup trucks, and semis. Virtually all of them have an electrical system, a transmission, a lighting system, tires, turn signals, a horn, etc. One could argue that because of the resemblances, the cars must have evolved from the motorcycles and the

pickup trucks from the cars—but, in reality, most of these vehicles are made by different factories and engineers scattered around the world. So why do they have so many similarities? It's because they operate in the same environment, and it makes sense for them to have comparable functions to work successfully.

Likewise, the creatures functioning in the same environment should have many similarities.

As we study the creatures of our planet, perhaps we must look more carefully at what they can tell us. In the ancient book of Job, a wise Old Testament man revealed how creation is a witness to God's power:

> But now ask the beasts, and they will teach you; and the birds of the air, and they will tell you; or speak to the earth, and it will teach you; and the fish of the sea will explain to you. Who among all these does not know that the hand of the Lord has done this, in whose hand is the life of every living thing, and the breath of all mankind? (12:7–10).

The similarities we see among living things in our world are best explained by creation, not evolutionary naturalism. As our knowledge in the sciences grows, the theory of evolutionary change does not hold up. Evidence for a long, gradual line of progressive change is missing. The relationships we observe do not indicate we evolved from a common ancestor 3.5 billion years ago, but instead, show we were created by a common Maker about 6,000 years ago.

More Circular Reasoning

In the context of homology, circular reasoning essentially says, "These structures are homologous because they have a common ancestor, and they have a common ancestor because they are homologous."

Radiocarbon Dating

An Amazing Fact:

Evolutionists claim that it took the Colorado River five million years to carve out the Grand Canyon. But following an eruption of Mount Saint Helens in 1982, a mudflow pushed the water in Spirit Lake surging over its borders. Rocks, mud, and logs churned 20 miles down the valley, carving a miniature Grand Canyon 140 feet deep, complete with layers and strata, in days!

You open a door and step into a four-walled room—it has a ceiling but no windows. The only thing in the room is a lit, upright candle on the floor. The candle is 10 inches high and 2 inches wide.

Here's the big question: How long has the candle been burning?

You can measure the current rate of the burn and the accumulating puddle of wax at the base. Maybe you can also measure the soot on the ceiling. But there are some things you just can't know. How tall was the candle when it started burning, and how wide was it at the top? Was the top composed of the same material and burning at the same speed as what you're measuring now? When you opened the door, you let in additional oxygen, changing the burn rate and scrambling your calculations. How many times has that door been opened anyway?

Without a first-hand witness to give you additional data, it's simply impossible to know how long the candle has been burning.

However, measuring the age of bones, rocks, and wood is even more challenging.

Much of evolutionary theory hinges upon conventional dating methods, but these methods rest on significant assumptions. If the dating methods are not accurate, the theory implodes.

For instance, you've likely heard that evolutionists use a method called carbon dating to determine the age of organic materials based on their carbon-14 content, a radioactive isotope of carbon. They believe that this method is effective for dating materials up to about 50,000 years old. However, many serious problems with carbon dating suggest it may not be as ironclad as the public has been led to believe. Here are just a few such issues:

1. **Carbon dating relies on unproven assumptions about the past.** Most scientists assume that the amount of carbon-14 in the atmosphere has remained constant over time. Yet evidence suggests that the amount of atmospheric carbon-14 has varied significantly over time, which would greatly affect the accuracy of results. For instance, volcanic activity ejects huge amounts of carbon into the air, greatly affecting carbon-14 ratios. Furthermore, the rate of decay of carbon-14 is

Continued on page 65.

Why So Unreliable?

When it comes to the shaky foundations of carbon dating, there are a few possible explanations for why we see inconsistencies. One of these is the earth's magnetic field. Studies have shown that changes in the magnetic field can affect the rate at which carbon-14 decays. This could explain why we sometimes see such a wide range of carbon dating results. Another possible explanation is that carbon dating can be impacted by contamination of the sample being tested. Even a small amount of contamination can greatly impact the results. In some cases, the presence of more recent carbon in the sample can also invalidate carbon dating results altogether. A global flood, exposing samples to sodium-saturated water for a year, would have also significantly contaminated sample readings.

Dating a Young Earth

Other dating methods, such as helium diffusion and radioisotope dating, suggest a young age for the earth. For instance, one experiment involved dating zircons, which are minerals, from granite rocks using both carbon and helium diffusion dating methods. The zircons were sampled from "Precambrian" granite in New Mexico and a mine in South Africa. The dating results suggested that the rocks were more than a billion years old. However, the helium diffusion dating results suggested that the rocks were only a few thousand years old, based on the rate at which helium atoms diffused out of the zircons. Evolutionary scientists disputed this latter result—but why? Not because the data was wrong. Rather, the data went against their assumptions!

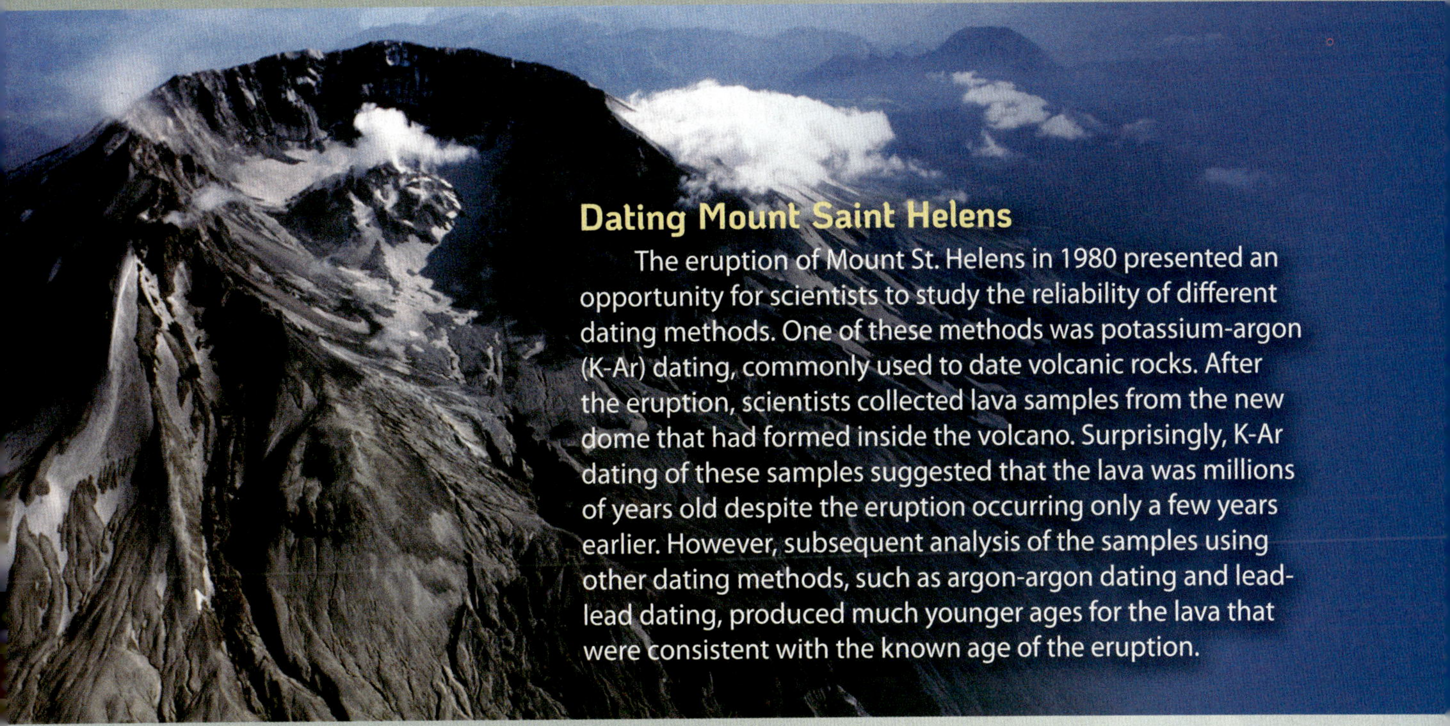

Dating Mount Saint Helens

The eruption of Mount St. Helens in 1980 presented an opportunity for scientists to study the reliability of different dating methods. One of these methods was potassium-argon (K-Ar) dating, commonly used to date volcanic rocks. After the eruption, scientists collected lava samples from the new dome that had formed inside the volcano. Surprisingly, K-Ar dating of these samples suggested that the lava was millions of years old despite the eruption occurring only a few years earlier. However, subsequent analysis of the samples using other dating methods, such as argon-argon dating and lead-lead dating, produced much younger ages for the lava that were consistent with the known age of the eruption.

assumed to have been constant, decaying at the same rate today as it did millennia ago—but unless we discover time travel, there is no way to test this.

2. **Carbon dating can produce vastly inconsistent results.** Carbon dating of living mollusk shells once showed that they were more than 2,000 years old, even though they were known to be less than 100 years old. The carbon dating of wood from a burial chamber of the First Dynasty in Egypt dated it to be nearly 7,500 years old, but other methods dated it to be only 1,000 years old. These inconsistencies are not limited to ancient materials. One study found that carbon dating of human tissue produced results that varied by as much as 3,500 years.

3. **Carbon dating becomes less reliable over time.** Carbon-14 has a half-life of about 5,700 years, which means that after that time, half of the original carbon-14 atoms in a sample would have decayed to nitrogen-14. As a result, the amount of carbon-14 in a sample decreases exponentially over time and becomes difficult to measure accurately beyond a certain point because the sample size is so small.

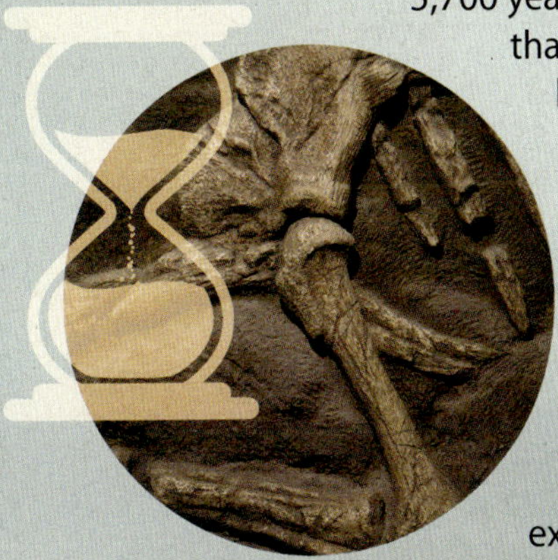

While carbon dating can be a useful research tool for scientists in the short term, these and other issues with carbon dating strongly suggest that the method is not reliable enough to conclusively demonstrate that organic matter is as old as evolutionists claim.

SOURCES

"Radiocarbon Dating and American Evangelical Christians," Perspectives on Science and Christian Faith

"The Pitfalls of Radiocarbon Dating," Answers in Genesis

"Radiocarbon Dating and Questions," Answers in Genesis

"Carbon Dating Undercuts Evolution's Long Ages," Creation Ministries International

"How Accurate Is Radiocarbon Dating?" NOVA

"Earth's magnetic field could change carbon dating results," New Scientist

"Carbon-14 Dating Does Not Disprove the Bible," Institute for Creation Research

"The Earth's Magnetic Field: Evidence That the Earth Is Young," Institute for Creation Research

"Helium Diffusion Rates Support Accelerated Nuclear Decay," Institute for Creation Research

"Radioisotope Dating of Grand Canyon Rocks: Another Devastating Failure for Long-Age Geology," Creation Ministries International

"The 1986 dacite lava dome at Mount St Helens volcano: how high was it?" Journal of Creation.

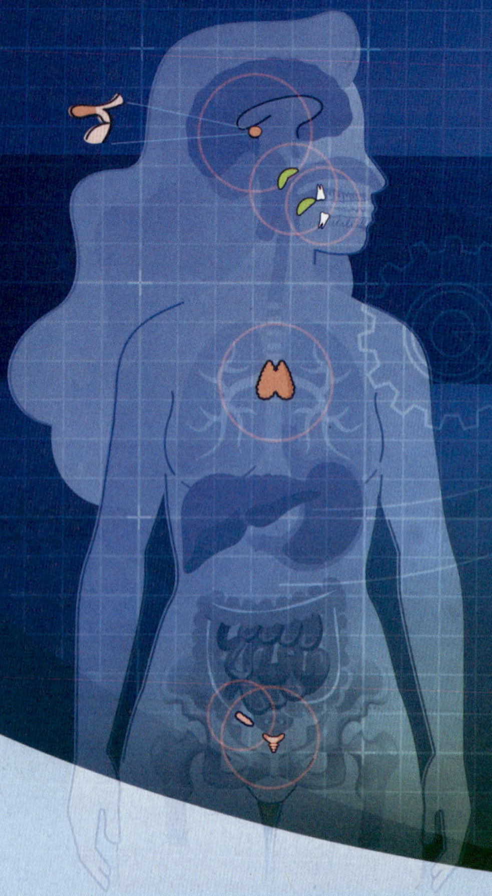

Vestigial Organs: Unnecessary Parts?

An Amazing Fact:

The coccyx—the tailbone—was once thought to be a "remnant tail" inherited from our apelike ancestors. Though some textbooks suggest the coccyx has no function, new evidence shows it aids in providing a muscular base that holds organs of the abdomen in place.

One of the arguments used to support evolutionary biology is the existence of so-called "vestigial organs" in the human body. Supposedly, these "useless parts" once served a purpose in our evolutionary past. It was argued that such organs must have been helpful to human survival—but over millions of years, they became unnecessary and lost their function. Some are used to show our supposed common ancestry with animals.

Evolutionists once declared nearly 180 organs as vestigial—including the appendix, tonsils, pineal gland, and thymus—useful in the evolutionary past but having no value today. But as the fields of anatomy and physiology have advanced, new discoveries have revealed that these body parts are actually functional and often critical to our survival!

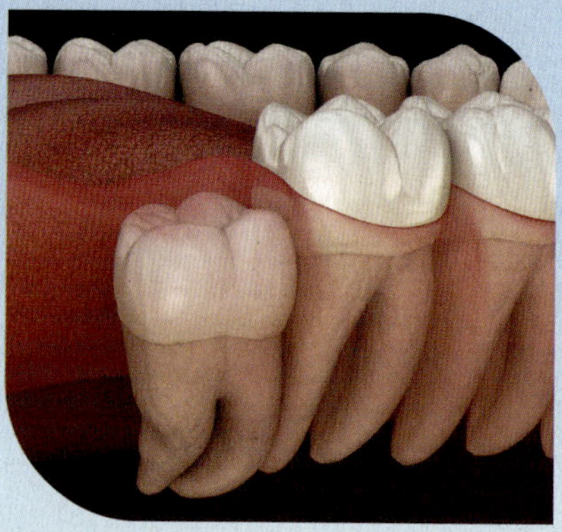

For example, the appendix was once considered an obsolete leftover of digestive evolution. It seemed to serve no purpose, and it could get infected and create problems, so why not get rid of it? But we now know it has a valuable immunological function as a front-line protector against infection, similar to the tonsils.

Wisdom teeth have been commonly touted as evidence for evolution. Many have these back molars removed as teenagers. Is this because God designed our mouths poorly—or are they evolutionary leftovers? The problem of impaction with wisdom teeth is a recent phenomenon, beginning around the time of the Industrial Revolution in the 1700s. Before this, few people had issues with wisdom teeth. What caused the change?

People used to eat much coarser food, which means they chewed harder. This strengthened and enlarged the jawbone, making a larger space for molars. Even today, people within the same country have different experiences with wisdom teeth. Rural residents who eat a more fibrous diet have few problems with their back molars as compared to their suburban counterparts, who eat "soft" modern food.

Now, experts recommend that if people's family genes provide enough room in their mouths, they should not have their wisdom teeth removed, for they are actually useful in chewing.

There is an infinitely wise purpose for all of the anatomy in all of God's creatures. "I will praise You, for I am fearfully and wonderfully made; marvelous are Your works, and that my soul knows very well" (Psalm 139:14).

Science and Religion:
In Conflict or Harmony?

An Amazing Fact:

Karl Popper, British-Austrian science philosopher and "one of the 20th century's most influential philosophers of science," said, "We can never give positive reasons which justify the belief that a theory is true. [Science is not] a system of certain, or well-established, statements; nor is it a system which steadily advances toward a state of finality. Our science is not knowledge: It can never claim to have attained truth, or even a substitute for it."

Science is often presented as being objective, dealing only with empirical data. Religion, including the Bible, is mostly seen as subjective, dealing with the supernatural and personal experience. However, history reveals a more nuanced picture.

Many early scientists were motivated by a desire to understand the divine. For instance, some of the most revered scientists in history, including Galileo, Newton, and Kepler, viewed their scientific inquiries as a means to uncover God's creation and laws. They believed that by studying the natural world, they could appreciate the intricacies of divine design.

The scientific revolution (1543–1687) actually emerged in a time of deeply rooted religious thought. Early scientific studies were mostly supported by religious institutions, which provided funding and encouragement for scholars to explore the mysteries of the universe. The quest for knowledge was seen as a sacred mission, and scientific discoveries were often framed as revelations of God's handiwork.

However, the rise of secular humanism during the Enlightenment (17th/18th centuries) changed this dynamic. Secular humanism emphasizes reason, ethics, and human values independent of religious considerations, advocating for a worldview grounded in **materialism.** As secular humanism gained traction, it led to the perception that the sciences stood in stark opposition to religious belief.

When Charles Darwin suggested that species could change over long ages—a belief opposing the Genesis creation account—many scientists were ready to embrace it to escape the church's restrictions. Over time, accepting materialism as an absolute became the foundation for scientific thinking.

But if science is a search for objective truth about our material world, wouldn't we want to consider all options for our origins? Wouldn't we want to follow the evidence wherever it may lead us?

Materialism

The philosophical belief that everything in the universe, including life and consciousness, can be explained solely by physical matter and its interactions, denying the need for a supernatural creator and supporting the idea that life arose through purely natural evolutionary processes.

Conclusion: Do You Want to Live With Purpose?

An Amazing Fact:

The second-leading cause of death among people aged 15 to 24 in the United States is suicide. In the 1950s, about three out of 100,000 teenagers committed suicide. In the 1960s, after public schools began teaching evolution as the preferred view on the origin of life, the average suicide rate of teenagers shot up 300 percent.

At the start of this magazine, we asked you to consider three questions:

1. **Where did I come from?**
2. **What am I doing here?**
3. **Where am I going?**

Before you dismiss these questions, consider the implications of your answers.

Depending on your worldview, your responses to these questions about origins, purpose, and the meaning of life will take you in one of two completely different directions.

Though there are many worldviews, they mostly fall into two categories: one, life evolved from lifeless elements (materialism); two, life was created by an intelligent designer (creationism).

In this magazine, we've considered an abundance of evidence for a Creator, described in the Bible as the One who made the earth and created life.

> In the beginning was the Word, and the Word was with God, and the Word was God. He was in the beginning with God. *All things were made through Him, and without Him nothing was made that was made*. In Him was life, and the life was the light of men (John 1:1–4, emphasis added).

The alternative view is that there is no God, no intelligent being who created our world. Evolution says that through a series of random accidents, our cosmos and world came from nothing and that life formed by chance. This would answer the first question, *Where did I come from?* Nothing.

With that in mind, how does the evolutionist answer the next question, *What am I doing here?* For them, life can't have a real purpose; indeed, even the concept of purpose is a mere accident of survival of the fittest. If coming into existence were a mere accident, how could anyone have a purpose?

Many dangerous and cruel consequences for our culture come with this evolutionary worldview of ultimate meaninglessness. Suppose we don't believe we were created to reflect the image of God—that

impacts our views on hot topics like abortion and euthanasia. Survival of the fittest says a developing human fetus or a disabled elderly person is weaker and, therefore, taking up resources best utilized by the strongest.

Worse, discrimination based on race has its foundation in evolution! The original title of Charles Darwin's most famous book reveals racism's connection to materialism: *On the Origin of Species by Means of Natural Selection, or the Preservation of Favored Races in the Struggle for Life.* Doesn't "favored races" imply "unfavored races"? Is it any wonder that Joseph Stalin and Adolf Hitler, two of history's most cruel practitioners of genocide, believed in evolution and Darwin's theory of superior races?

The Gorilla in the Room

A looming conundrum hangs over the discussion regarding creation versus evolution. Clearly, many highly intelligent people oppose intelligent design.

Why do so many well-educated people cling to the theory of evolution while ignoring the implications of proven laws—such as the law of biogenesis, which says that life comes only from pre-existing life? Or the law of conservation of mass, which says that matter does not create itself?

Put simply, they are choosing to believe that everything came from nothing and for no purpose. But why would anyone embrace such a dismal worldview?

Perhaps one reason is that if there is a super-intelligent being

who created life, then He would have ultimate authority over that life. It would also stand to reason that if He made His intelligent creatures with a purpose, then disregarding that purpose would have consequences.

Scripture says, "We must all appear before the judgment seat of Christ, that each one may receive the things done in the body, according to what he has done, whether good or bad" (2 Corinthians 5:10), and that "each of us shall give account of himself to God" (Romans 14:12). In a sense, believing in evolution, that humans are just evolved animals, provides a convenient excuse for living a life free from moral constraints and accountability.

Yet, is it really worth sacrificing eternal life in a new world untainted by sin, death, and disease for a short life of sin on Earth?

Discovering Your Identity

It's perfectly rational to question humanity's origin—after all, understanding our past shapes our identity today. If you don't know your history, you don't fully know who you are today. Thousands of self-help books are written to help people understand their identity. However, most are based on a materialistic, evolutionary view that ultimately offers no deep purpose or a future with any real meaning.

Indeed, perhaps we have been prewired by a loving Creator to passionately search for meaning because He knows we will never be satisfied unless we have a loving relationship with One greater than ourselves. Maybe this longing for something more is a call to reconsider the Bible's view of a caring God who originally created a perfect world.

While evolution teaches that death is simply the natural process that leads to more highly evolved life forms, the Bible describes death as an enemy (1 Corinthians 15:26). Evolution cannot explain why our bodies have the

Theistic Evolution: Straddling the Fence

Some people try to straddle the fence between the opposing worldviews of creation and evolution. These people don't deny the evidence that a superintelligence is behind the miracle of life—but they struggle to dismiss the many scientists who advocate for evolution. They occupy a supposed middle ground, arguing that God used evolution to create. This is known as "theistic evolution," a worldview endorsed by the Vatican and many Protestant universities.

However, this belief is incompatible with the biblical account! For one, Jesus taught that God made humanity in one day, not through eons of trial and error and random mutations. Some say the first 11 chapters of Genesis aren't meant to be taken literally, but Jesus viewed them as literal historical accounts (Matthew 19:4, 5; 23:35; 24:37–39). He also stated, "If you believed Moses, you would believe Me; for he wrote about Me. But if you do not believe his writings, how will you believe My words?" (John 5:46, 47). To reject Genesis as literal history is to suggest Jesus was mistaken.

Theistic evolution teaches that by God's design, life on Earth began with millions of years of pain and death, with species killing and devouring one another. But the Bible says God created the world as a perfect, peaceful paradise and that death and pain entered only as a result of human sin.

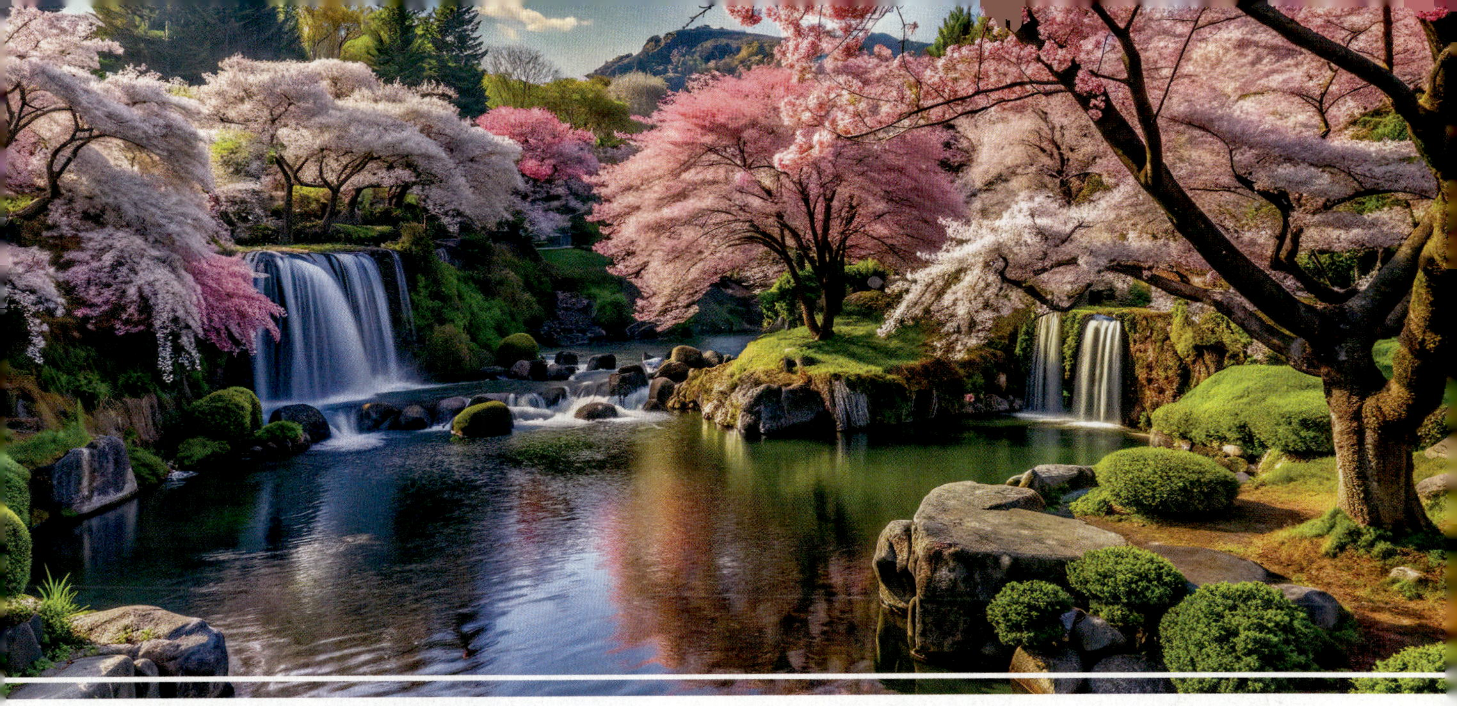

stunning ability to heal and fight disease for most of our lives, but then mysteriously lose that capacity with age. A lizard can grow a new tail, and an octopus can grow a new tentacle. Some jellyfish might live indefinitely if not eaten or injured. So why do humans grow old and die?

Why Suffering?

Suffering and death were not a part of God's original design. His plan for humans was to live forever in a breathtaking paradise.

It's not on a scientific basis that many have turned away from believing in God. Instead, they see the evil, suffering, and death in our world and reason that if there were an all-powerful and loving God, He would immediately put an end to these atrocities.

But there is a reasonable explanation—and it's found in the truth about how sin began and about God's plan to end evil, suffering, and death.

The Bible says, "In the beginning God created the heavens and the earth" (Genesis 1:1). When He finished making all things, including humans, He "saw everything that He had made, and indeed it was very good" (v. 31). In the beginning, there was no evil, suffering, or death on Earth!

The Bible also teaches that "God is love" (1 John 4:8), and it is the nature of love to give genuine freedom. So instead of making humans to be like mere robots who are preprogrammed to obey, God created us with free will.

That's also why Lucifer, a rebellious angel, was allowed to disobey God. Eventually, Lucifer, now known as Satan, was restricted to our planet, where he continued his campaign against God (Revelation 12:7–9). Satan's first act on Earth was to entice the first humans, Adam and Eve, to distrust and disobey their Creator—cutting them off from the source of life and opening the door to suffering and death. Through their disobedience, evil and death entered our world.

But God did not abandon our fallen planet! To win back the hearts of humans and free them from the grip of sin, Jesus came to Earth. He freely chose to die for our sins, exchanging His life for ours (2 Corinthians 5:21). Jesus invites us to humbly confess our sins to God, receive forgiveness, and be empowered to live a new and better life.

The Son of God has purchased for us the opportunity to be freed from the guilt and grip of sin and to live in eternal happiness (1 Peter 1:18, 19). This is called the gospel—the "good news." It is a gift offered to everyone in our world. It is available to you right now. "The wages of sin is death, but the gift of

God is eternal life in Christ Jesus our Lord" (Romans 6:23). Will you receive it?

There is a war raging between God and Satan; it has been called "the great controversy between good and evil." Each one of us must choose a side. God has allowed the devil to promote his ways so that we can see the results of disregarding the Creator's life-giving laws.

Evolutionists and creationists alike can see that the world is careening out of control—massive overconsumption of resources, destruction of the environment, and global conflicts with devastating weapons plague the globe. Soon, God will intervene to end Satan's campaign, and Christ will come to rescue the redeemed. The living today still have a chance to choose between God's way of love, which leads to life, and Satan's way of pride and selfishness, which leads to death.

Ultimately, everyone will conclude that God is not cruel or unfair (Philippians 2:10, 11; Revelation 15:3). Justice will prevail (20:11, 12). God will destroy evil and death (21:4) and restore the earth to its original, perfect design (v. 1). The saved will know forever that God is love. Every heart will be at peace and secure in the knowledge that the Creator can be trusted.

As a result, death, cruelty, sorrow, and pain will end—and sin will never rise again (Nahum 1:9).

God's True Character

Many are unaware of what the Bible says about God's character because they have not studied it for themselves. Our Creator is described as a compassionate Father who is not apathetic toward this rebellious planet. He is "The Lord, the Lord God, merciful and gracious, longsuffering, and abounding in goodness and truth" (Exodus 34:6). He is "not willing that any should perish but that all should come to repentance" (2 Peter 3:9).

Though sin has marred our earth, we still see God's love written in nature's beautiful and marvelous design. "He loves righteousness and justice; the earth is full of the goodness of the

Marriage and Gender in a Purposeless Life

Evolutionary philosophy rejects the sanctity of marriage between one man and one woman, an institution with foundations in the Bible, and it undermines biological reality in favor of pop-culture philosophy. "From the beginning of the creation, God 'made them male and female.' 'For this reason a man shall leave his father and mother and be joined to his wife, and the two shall become one flesh'" (Mark 10:6–8).

LORD" (Psalm 33:5). When we study the things in nature, we see that they harmonize with the Bible.

But His character is most fully revealed in the life of Jesus Christ. "He who has seen Me [Jesus] has seen the Father" (John 14:9). The self-sacrificing love of God was fully revealed when Christ chose to die on the cross for our sins.

> For God so loved the world that He gave His only begotten Son, that whoever believes in Him should not perish but have everlasting life. For God did not send His Son into the world to condemn the world, but that the world through Him might be saved (John 3:16, 17).

Finally, through studying our beginnings and learning about the character of God, we can know the answer to the third question, *Where am I going?* The Bible reveals that God has a purpose for your life. "I know the plans I have for you, declares the LORD, plans for welfare and not for evil, to give you a future and a hope" (Jeremiah 29:11 ESV). He freely offers a future of happiness and eternal life to all who humbly come to Him. "He who hears My word and believes in Him who sent Me has everlasting life, and shall not come into judgment, but has passed from death into life" (John 5:24).

Because God is love, and love cannot be forced, He gives us freedom to either follow or reject Him. "Choose for yourselves this day whom you will serve" (Joshua 24:15).

Right now, God is calling you to believe in Him as your Maker. "Come to Me, all you who labor and are heavy laden, and I will give you rest" (Matthew 11:28).

Will you now take your Creator's outstretched hand—the hand that formed the world?

A Reminder of Creation

Did you know that God gave humanity a weekly reminder of His creative power? On the seventh day of creation week, "God ended His work which He had done, and He rested on the seventh day from all His work which He had done. Then God blessed the seventh day and sanctified it, because in it He rested from all His work which God had created and made" (Genesis 2:2, 3).

All the world recognizes this seven-day week, but nothing in astronomy gives us this weekly cycle; it can be traced only to the biblical account of creation. In more than 30 major languages of the world, the word for the seventh day of the week is Sabbath. In Spanish, it's *sábado*; in Russian, it's *subota*—indicating the seventh day was memorialized in antiquity as a sacred reminder of creation.

Jesus said, "The Sabbath was made for man" (Mark 2:27). The word he used for "man" is *anthropos,* which means the human race—everybody, not just Jewish people. And to keep us from forgetting the Creator and how we came into being, the fourth commandment says, "Remember the Sabbath day, to keep it holy" (Exodus 20:8). Why? "For in six days the LORD made the heavens and the earth, the sea, and all that is in them, and rested the seventh day. Therefore the LORD blessed the Sabbath day and hallowed it" (v. 11).

Learn more about how the war between good and evil began and will end by reading our magazine *Cosmic Conflict: The Origin of Evil.* See page 74 for details.

Did you enjoy this magazine?

Then you'll LOVE the rest of our truth-filled, biblical magazines that will open your eyes with in-depth, fast-paced teaching and eye-catching, colorful graphics.

Only $3.95 each!

Cosmic Conflict

The Origin of Evil

The war between good and evil is a mystery, but the Bible reveals its origin and impact on your life! This powerful magazine explains God's true character and His plan to redeem humanity.

BK-CCOE

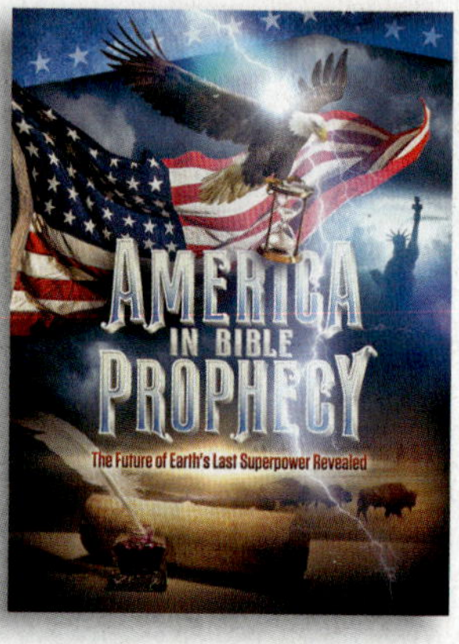

America in Bible Prophecy

The Future of Earth's Last Superpower Revealed

The book of Revelation describes two end-time powers on Earth that will persecute God's people before Jesus returns. This eye-opening magazine shares the key Bible facts that will amaze and inspire you!

BK-AIP

Earth's Final Warning

The Three Angels of Revelation

Explore the vital end-time messages from the three angels of Revelation 14. Packed with Scripture and fascinating facts, this colorful magazine will bring you peace and hope!

BK-3AMR

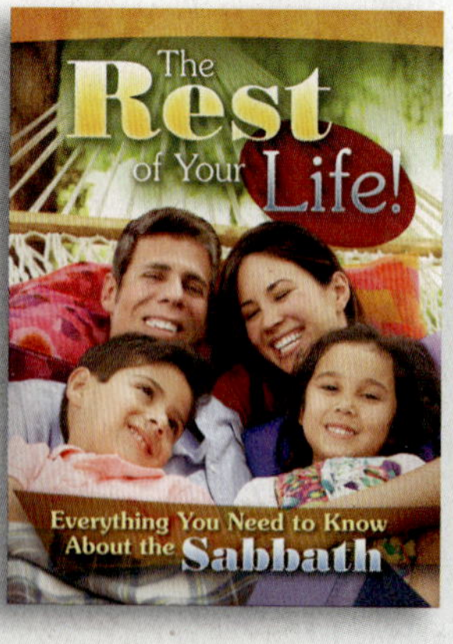

The Rest of Your Life!

Everything You Need to Know About the Sabbath

Reveals the amazing truth about the seventh-day Sabbath, from ancient times to the present, and what it means for you!

BK-RYL

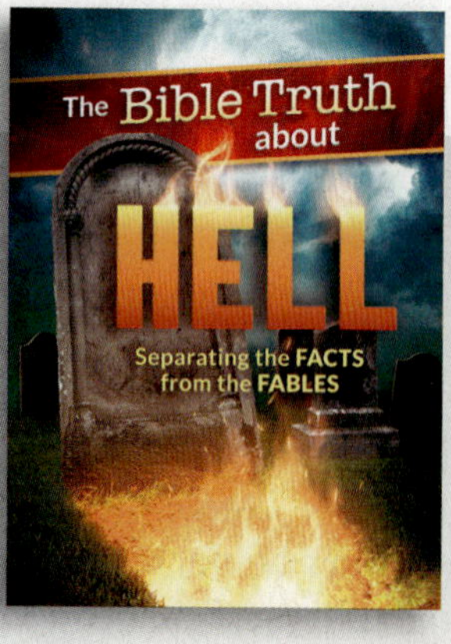

The Bible Truth About Hell

Separating the Facts from the Fables

Is hell a real place—or just a metaphor? Are sinners tortured in hellfire for eternity? Is the devil in charge of it all? These questions and more are answered in this colorful magazine.

BK-BTH